Student Solutions Manual

to accompany

Elementary Statistics

A Step By Step Approach

Eighth Edition

Allan Bluman

Community College of Allegheny Country

Prepared by

Sally Robinson

South Plains College

The McGraw-Hill Companies

Mc Graw Hill) Connect Learn Succeed™

Student Solutions Manual to accompany
ELEMENTARY STATISTICS: A STEP BY STEP APPROACH, EIGHTH EDITION
ALLAN BLUMAN

Published by McGraw-Hill Higher Education, an imprint of The McGraw-Hill Companies, Inc., 1221 Avenue of the Americas, New York, NY 10020. Copyright © 2012, 2009, 2007 and 2004 by The McGraw-Hill Companies, Inc. All rights reserved.

This book is printed on acid-free paper.

1 2 3 4 5 6 7 8 9 0 QDB/QDB 1 0 9 8 7 6 5 4 3 2 1

ISBN: 978–0–07–743865–4
MHID: 0–07–743865–5

Contents

Solutions to the Exercises

Preface

This manual includes solutions to odd and selected even exercises in *Elementary Statistics: A Step-by-Step Approach 8e* by Allan G. Bluman. Solutions are worked out step-by-step where appropriate and generally follow the same procedures used in the examples in the textbook. Answers may be carried to several decimal places to increase accuracy and to facilitate checking. See your instructor for specific rounding rules. Graphs are included with the solutions when appropriate or required. They are intended to convey a general idea and may not be to scale.

Caution: Answers generated using graphing calculators such as the TI-83 may vary from those shown in this manual.

To maximize the assistance provided in this manual, you should:

1. Read each section of the text carefully, noting examples and formulas.

2. Begin working the exercises using textbook examples and class notes as your guide, then refer to the answers in this manual.

3. Many instructors require students to interpret their answers within the context of the problem. These interpretations are powerful tools to understanding the meaning and purpose of each calculation. You should attempt to interpret each calculation even if you are not required to do so.

4. Be sure to show your work. When checking your work for errors you will need to review each step. When preparing for exams, reviewing each step helps you to recall the process involved in producing each calculation.

5. As you gain confidence and understanding, you should attempt to work exercises without referring to examples or notes. Check each answer in the solutions manual before beginning the next exercise.

6. Slight variations between your answers and the answers in this manual are probably due to rounding differences and should not be a cause for concern. If you are concerned about these variations, check each step of your calculation again.

7. Many errors can be traced to the improper application of the rules for order of operations. You should first attempt to determine where and how your error occurred because diagnosing your error increases understanding and prevents future errors. See your instructor if you are unsure of the location or cause of your error.

Sally H. Robinson

REVIEW EXERCISES - CHAPTER 1

1. Descriptive statistics describes a set of data. Inferential statistics uses a set of data to make predictions about a population.

3. Answers will vary.

5. When the population is large, the researcher saves time and money using samples. Samples are used when the units must be destroyed.

6.
 a. Inferential   e. Inferential
 b. Descriptive   f. Inferential
 c. Descriptive   g. Descriptive
 d. Descriptive   h. Inferential

7.
 a. Ratio   f. Ratio
 b. Ordinal   g. Ordinal
 c. Interval   h. Ratio
 d. Ratio   i. Ratio
 e. Ratio   j. Nominal

8.
 a. Qualitative   e. Quantitative
 b. Quantitative   f. Quantitative
 c. Quantitative   g. Quantitative
 d. Qualitative

9.
 a. Discrete   e. Continuous
 b. Continuous   f. Discrete
 c. Discrete   g. Continuous
 d. Continuous

11. Random samples are selected by using chance methods or random numbers. Systematic samples are selected by numbering each subject and selecting every kth number. Stratified samples are selected by dividing the population into groups and selecting from each group. Cluster samples are selected by using intact groups called clusters.

12.
 a. Cluster   d. Systematic
 b. Systematic   e. Stratified
 c. Random

13. Answers will vary.

15. Answers will vary.

17.
 a. Experimental   c. Observational
 b. Observational   d. Experimental

19. Answers will vary. Possible answers include:
(a) overall health of participants, amount of exposure to infected individuals through the workplace or home
(b) gender and/or age of driver, time of day
(c) diet, general health, heredity factors
(d) amount of exercise, heredity factors

21. Claims can be proven only if the entire population is used.

23. Since the results are not typical, the advertisers selected only a few people for whom the product worked extremely well.

25. "74% more calories" than what? No comparison group is stated.

27. What is meant by "24 hours of acid control"?

29. Possible reasons for conflicting results: The amount of caffeine in the coffee or tea or the brewing method.

31. Answers will vary.

CHAPTER QUIZ
 1. True
 2. True
 3. True
 4. False, it is stratified sampling.
 5. True
 6. True
 7. False, it is 5.5-6.5 inches.
 8. c
 9. b
 10. d
 11. a
 12. c
 13. a
 14. Descriptive, inferential
 15. Gambling, insurance
 16. Population
 17. Sample

18.
a. Saves time
b. Saves money
c. Use when population is infinite

19.
a. Random   c. Cluster
b. Systematic d. Stratified

20. Quasi-experimental

21. Random

22.
a. Descriptive d. Inferential
b. Inferential   e. Inferential
c. Descriptive

23.
a. Nominal d. Interval
b. Ratio   e. Ratio
c. Ordinal

24.
a. Continuous d. Continuous
b. Discrete   e. Discrete
c. Continuous

25.
a. $31.5 - 32.5$ minutes   d. $18.5 - 19.5$ pounds
b. $0.475 - 0.485$ millimeter e. $12.05 - 12.15$ quarts
c. $6.15 - 6.25$ inches

EXERCISE SET 2-1

1. Frequency distributions are used to organize data in a meaningful way, to facilitate computational procedures for statistics, to make it easier to draw charts and graphs, and to make comparisons among different sets of data.

3.
a. $31.5 - 38.5$, $\frac{32+38}{2} = \frac{70}{2} = 35$, $38.5 - 31.5 = 7$
b. $85.5 - 104.5$, $\frac{86+104}{2} = \frac{190}{2} = 95$, $104.5 - 85.5 = 19$
c. $894.5 - 905.5$, $\frac{895+905}{2} = \frac{1800}{2} = 900$, $905.5 - 894.5 = 11$
d. $12.25 - 13.55$, $\frac{12.3+13.5}{2} = \frac{25.8}{2} = 12.9$, $13.55 - 12.25 = 1.3$
e. $3.175 - 4.965$, $\frac{3.18+4.96}{2} = \frac{8.14}{2} = 4.07$, $4.965 - 3.175 = 1.79$

5.
a. Class width is not uniform.
b. Class limits overlap, and class width is not uniform.
c. A class has been omitted.
d. Class width is not uniform.

7.

Class	Tally	f	Percent
A	IIII	4	10%
M	HI HI HI HI HI III	28	70%
H	HI I	6	15%
S	II	2	5%
		40	100%

9. $H = 325$ $L = 165$
Range $= 325 - 165 = 160$
Width $= 160 \div 8 = 20$ round up to 21

Limits	Boundaries	f
165 - 185	164.5 - 185.5	4
186 - 206	185.5 - 206.5	6
207 - 227	206.5 - 227.5	15
228 - 248	227.5 - 248.5	13
249 - 269	248.5 - 269.5	9
270 - 290	269.5 - 290.5	1
291 - 311	290.5 - 311.5	1
312 - 332	311.5 - 332.5	1
		50

9. continued
A peak occurs in class 207 - 227. There are no empty classes. Each of the three highest classes has one data value.

	cf
Less than 164.5	0
Less than 185.5	4
Less than 206.5	10
Less than 227.5	25
Less than 248.5	38
Less than 269.5	47
Less than 290.5	48
Less than 311.5	49
Less than 332.5	50

11. $H = 780$ $L = 746$
Range $= 780 - 746 = 34$
Width $= 34 \div 5 = 6.8$ round up to 7

Limits	Boundaries	f
746 - 752	745.5 - 752.5	4
753 - 759	752.5 - 759.5	6
760 - 766	759.5 - 766.5	8
767 - 773	766.5 - 773.5	9
774 - 780	773.5 - 780.5	3
		30

	cf
Less than 745.5	0
Less than 752.5	4
Less than 759.5	10
Less than 766.5	18
Less than 773.5	27
Less than 780.5	30

13. $H = 70$ $L = 27$
Range $= 70 - 27 = 43$
Width $= 43 \div 7 = 6.1$ or 7

Limits	Boundaries	f
27 - 33	26.5 - 33.5	7
34 - 40	33.5 - 40.5	14
41 - 47	40.5 - 47.5	15
48 - 54	47.5 - 54.5	11
55 - 61	54.5 - 61.5	3
62 - 68	61.5 - 68.5	3
69 - 75	68.5 - 75.5	2
		55

13. continued

	cf
Less than 26.5	0
Less than 33.5	7
Less than 40.5	21
Less than 47.5	36
Less than 54.5	47
Less than 61.5	50
Less than 68.5	53
Less than 75.5	55

15. $H = 635 \quad L = 6$
Width $= 635 - 6 = 629$
Range $= 629 \div 5 = 125.8$ round up to 127

Limits	Boundaries	f
6 - 132	5.5 - 132.5	16
133 - 259	132.5 - 259.5	3
260 - 386	259.5 - 386.5	0
387 - 513	386.5 - 513.5	0
514 - 640	513.5 - 640.5	1
		20

The greatest concentration of data values is in the lowest class. All but one of the data values are in the lowest two classes. There is one extremely large data value occurring in the highest class.

	cf
Less than 5.5	0
Less than 132.5	16
Less than 259.5	19
Less than 386.5	19
Less than 513.5	19
Less than 640.5	20

17. $H = 123 \quad L = 77$
Range $= 123 - 77 = 46$
Width $= 46 \div 7 = 6.6$ or 7

Limits	Boundaries	f
77 - 83	76.5 - 83.5	1
84 - 90	83.5- 90.5	1
91 - 97	90.5- 97.5	6
98 - 104	97.5- 104.5	14
105 - 111	104.5 - 111.5	8
112 - 118	111.5 - 118.5	1
119 - 125	118.5 - 125.5	1
		32

17. continued

	cf
Less than 76.5	0
Less than 83.5	1
Less than 90.5	2
Less than 97.5	8
Less than 104.5	22
Less than 111.5	30
Less than 118.5	31
Less than 125.5	32

19. The percents add up to 101%. They should total 100% unless rounding was used.

EXERCISE SET 2-2

1.

Eighty applicants do not need to enroll in the summer programs.

3.

Limits	Boundaries	f
3 - 45	2.5 - 45.5	19
46 - 88	45.5 - 88.5	19
89 - 131	88.5 - 131.5	10
132 - 174	131.5 - 174.5	1
175 - 217	174.5 - 217.5	0
218 - 260	217.5 - 260.5	1
		50

3. continued

	cf
Less than 2.5	0
Less than 45.5	19
Less than 88.5	38
Less than 131.5	48
Less than 174.5	49
Less than 217.5	49
Less than 260.5	50

The distribution is positively skewed.

5. continued

	cf
Less than 0.5	0
Less than 43.5	24
Less than 86.5	41
Less than 129.5	44
Less than 172.5	48
Less than 215.5	49
Less than 258.5	49
Less than 301.5	49
Less than 344.5	50

The distribution is positively skewed.

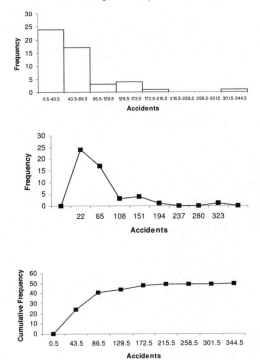

5.

Limits	Boundaries	f
1 - 43	0.5 - 43.5	24
44 - 86	43.5 - 86.5	17
87 - 129	86.5 - 129.5	3
130 - 172	129.5 - 172.5	4
173 - 215	172.5 - 215.5	1
216 - 258	215.5 - 258.5	0
259 - 301	258.5 - 301.5	0
302 - 344	301.5 - 344.5	1
		50

7.

Limits	Boundaries	f (1998)	f (2003)
0 - 22	-0.5 - 22.5	18	26
23 - 45	22.5 - 45.5	7	1
46 - 68	45.5 - 68.5	3	0
69 - 91	68.5 - 91.5	1	1
92 - 114	91.5 - 114.5	1	0
115 - 137	114.5 - 137.5	0	1
138 - 160	137.5 - 160.5	0	1
		30	30

7. continued

9. continued

Both distributions are positively skewed, but the data are somewhat more spread out in the first three classes in 1998 than in 2003. There are two large data values in the 2003 data.

9.

Limits	Boundaries	f
83.1 - 90.0	83.05 - 90.05	3
90.1 - 97.0	90.05 - 97.05	5
97.1 - 104.0	97.05 - 104.05	6
104.1 - 111.0	104.05 - 111.05	7
111.1 - 118.0	111.05 - 118.05	3
118.1 - 125.0	118.05 - 125.05	1
		25

	cf
Less than 83.05	0
Less than 90.05	3
Less than 97.05	8
Less than 104.05	14
Less than 111.05	21
Less than 118.05	24
Less than 125.05	25

11.

Limits	Boundaries	f
140 - 230	139.5 - 229.5	11
231 - 321	230.5 - 321.5	5
322 - 412	321.5 - 412.5	4
413 - 503	412.5 - 503.5	4
504 - 594	503.5 - 594.5	4
595 - 685	594.5 - 685.5	1
686 - 776	685.5 - 776.5	0
777 - 867	776.5 - 867.5	1
		30

	cf
Less than 139.5	0
Less than 229.5	11
Less than 321.5	16
Less than 412.5	20
Less than 503.5	24
Less than 594.5	28
Less than 685.5	29
Less than 776.5	29
Less than 867.5	30

The distribution is positively skewed.

11. continued

15.

$H = 270 \quad L = 80$

$Range = 270 - 80 = 190$

$Width = 190 \div 7 = 27.1 \text{ or } 28$

Use width $= 29$ (rule 2)

Limits	Boundaries	f	rf
80 - 108	79.5 - 108.5	8	0.17
109 - 137	108.5 - 137.5	13	0.28
138 - 166	137.5 - 166.5	2	0.04
167 - 195	166.5 - 195.5	9	0.20
196 - 224	195.5 - 224.5	10	0.22
225 - 253	224.5 - 253.5	2	0.04
254 - 282	253.5 - 282.5	2	0.04
			0.99*

*due to rounding

	crf
Less than 79.5	0.00
Less than 108.5	0.17
Less than 137.5	0.45
Less than 166.5	0.49
Less than 195.5	0.69
Less than 224.5	0.91
Less than 253.5	0.95
Less than 282.5	0.99

13.

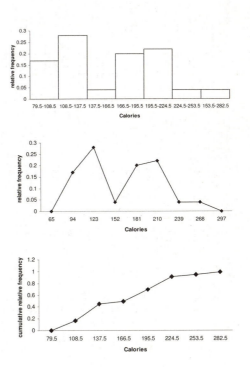

The histogram has two peaks.

The proportion of applicants who need to enroll in a summer program is 0.26 or 26%.

17.

Boundaries	crf
-0.5 - 27.5	0.87
27.5 - 55.5	0.03
55.5 - 83.5	0.00
83.5 - 111.5	0.03
111.5 - 139.5	0.00
139.5 - 167.5	0.03
167.5 - 195.5	0.03
	0.99

	crf
Less than -0.5	0.00
Less than 27.5	0.87
Less than 55.5	0.90
Less than 83.5	0.90
Less than 111.5	0.93
Less than 139.5	0.93
Less than 167.5	0.96
Less than 195.5	0.99

19.

Limits	Boundaries	X_m	f
22 - 24	21.5 - 24.5	23	1
25 - 27	24.5 - 27.5	26	3
28 - 30	27.5 - 30.5	29	0
31 - 33	30.5 - 33.5	32	6
34 - 36	33.5 - 36.5	35	5
37 - 39	36.5 - 39.5	38	3
40 - 42	39.5 - 42.5	41	2
			20

	cf
Less than 21.5	0
Less than 24.5	1
Less than 27.5	4
Less than 30.5	4
Less than 33.5	10
Less than 36.5	15
Less than 39.5	18
Less than 42.5	20

EXERCISE SET 2-3

1.

	f
May	18
June	79
July	101
August	344
September	459
October	280
November	61

1. continued

3.

5.

7.

9.

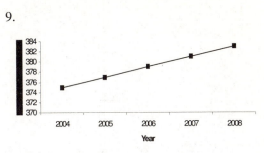

The atmospheric concentration of carbon dioxide has been steadily increasing over the years.

11.

Marital Status

Educational Attainment

- Less than 9th grade
- 9 - 12 but no diploma
- H.S. graduate
- Some college
- Bachelor's/advanced degree

13.

Career change	34%	122.4°
New job	29%	104.4°
Start business	21%	75.6°
Retire	16%	57.6°
	100%	360.0°

Pie chart:

13. continued
Pareto chart:

The pie graph better represents the data since we are looking at parts of a whole.

15.

4	2 3
4	6 6 7 8 9 9
5	0 1 1 1 1 2 2 4 4 4 4 4
5	5 5 5 5 6 6 6 7 7 7 7 8
6	0 1 1 1 2 4 4
6	5 8 9

The distribution is somewhat symmetric and unimodal. The majority of the Presidents were in their 50's when inaugurated.

17.

```
        Variety 1                        Variety 2
                      2 | 1 | 3  8
                    3 0 | 2 | 5
              9 8 8 5 2 | 3 | 6  8
                  3 3 1 | 4 | 1  2  5  5
      9 9 8 5 3 3 3 2 1 0 | 5 | 0  3  5  5  6  7  9
                          | 6 | 2  2
```

The distributions are similar but variety 2 seems to be more variable than variety 1.

19.
Answers will vary.

21.

In 1950, veal production was considerably higher than lamb. By 1970, production was approximately the same for both.

23.

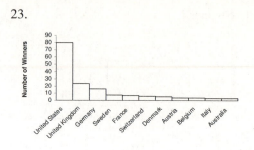

25. The values on the y axis start at 3.5. Also there are no data values shown for the years 2004 through 2011.

REVIEW EXERCISES - CHAPTER 2

1.

Class	f
Newspaper	10
Television	16
Radio	12
Internet	12
	50

3.

Class	f
Baseballs	4
Golf balls	5
Tennis balls	6
Soccer balls	5
Footballs	5
	25

5.

Class	f		cf
11	1	less than 10.5	0
12	2	less than 11.5	1
13	2	less than 12.5	3
14	2	less than 13.5	5
15	1	less than 14.5	7
16	2	less than 15.5	8
17	4	less than 16.5	10
18	2	less than 17.5	14
19	2	less than 18.5	16
20	1	less than 19.5	18
21	0	less than 20.5	19
22	1	less than 21.5	19
	20	less than 22.5	20

7.

Limits	Boundaries	f
15 - 19	14.5 - 19.5	3
20 - 24	19.5 - 24.5	18
25 - 29	24.5 - 29.5	18
30 - 34	29.5 - 34.5	8
35 - 39	34.5 - 39.5	3
		50

	cf
Less than 14.5	0
Less than 19.5	3
Less than 24.5	21
Less than 29.5	39
Less than 34.5	47
Less than 39.5	50

9.

Limits	Boundaries	f		cf
170 - 188	169.5 - 188.5	11	less than 169.5	0
189 - 207	188.5 - 207.5	9	less than 188.5	11
208 - 226	207.5 - 226.5	4	less than 207.5	20
227 - 245	226.5 - 245.5	5	less than 226.5	24
246 - 264	245.5 - 264.5	0	less than 245.5	29
265 - 283	264.5 - 283.5	0	less than 264.5	29
284 - 302	283.5 - 302.5	0	less than 283.5	29
303 - 321	302.5 - 321.5	1	less than 302.5	29
		30	less than 321.5	30

11.

Limits	Boundaries	rf
51 - 59	50.5 - 59.5	0.125
60 - 68	59.5 - 68.5	0.300
69 - 77	68.5 - 77.5	0.275
78 - 86	77.5 - 86.5	0.200
87 - 95	86.5 - 95.5	0.050
96 - 104	95.5 - 104.5	0.050
		1.000

	crf
Less than 50.5	0.000
Less than 59.5	0.125
Less than 68.5	0.425
Less than 77.5	0.700
Less than 86.5	0.900
Less than 95.5	0.950
Less than 104.5	1.000

11. continued

13.

15.

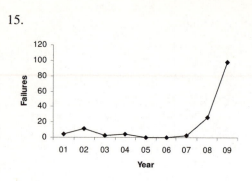

Bank failures increased in 2002 from 4 to 11, then dropped until 2008, when they increased to 28. The year 2009 brought an increase to 98.

17.

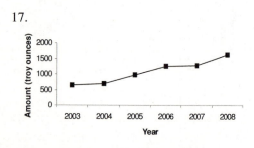

There has been a steady increase in the amount of gold produced by Columbia over the recent years.

19.

The majority of people surveyed would like to spend the rest of their careers with their present employer.

21.

10	2	8	8
11	3		
12			
13			
14	2	4	
15			
16			
17	6	6	6
18	4	9	
19	2		
20	5	9	
21	0		

CHAPTER 2 QUIZ

1. False
2. False
3. False
4. True
5. True
6. False
7. False
8. c
9. c
10. b
11. b
12. Categorical, ungrouped, grouped
13. 5, 20
14. Categorical
15. Time series
16. Stem and leaf plot
17. Vertical or y
18.

Class	f	cf
H	6	6
A	5	11
M	6	17
C	8	25
	25	

19.

20.

Class	f
0.5 − 1.5	1
1.5 − 2.5	5
2.5 − 3.5	3
3.5 − 4.5	4
4.5 − 5.5	2
5.5 − 6.5	6
6.5 − 7.5	2
7.5 − 8.5	3
8.5 − 9.5	4
	30

	cf
less than 0.5	0
less than 1.5	1
less than 2.5	6
less than 3.5	9
less than 4.5	13
less than 5.5	15
less than 6.5	21
less than 7.5	23
less than 8.5	26
less than 9.5	30

21.

22.

Limits	Boundaries	f
27 - 90	26.5 - 90.5	13
91 - 154	90.5 - 154.5	2
155 - 218	154.5 - 218.5	0
219 - 282	218.5 - 282.5	5
283 - 346	282.5 - 346.5	0
347 - 410	346.5 - 410.5	2
411 - 474	410.5 - 474.5	0
475 - 538	474.5 - 538.5	1
539 - 602	538.5 - 602.5	2
		25

23.

The distribution is positively skewed with one more than half of the data values in the lowest class.

24.

1	5	9			
2	6	8			
3	1	5	8	8	9
4	1	7	8		
5	3	3	4		
6	2	3	7	8	
7	6	9			
8	6	8	9		
9	8				

25.

26.

Note: Answers may vary due to rounding, TI 83's, or computer programs.

EXERCISE SET 3-1

1.
a. $\overline{X} = \frac{\Sigma X}{n} = \frac{93.09}{25} = 3.7236 \approx 3.724$

b. MD: 3.57, 3.64, 3.64, 3.65, 3.66, 3.67, 3.67, 3.68, 3.7, 3.7, 3.7, 3.73, **3.73**, 3.74, 3.74, 3.74, 3.75, 3.76, 3.77, 3.78, 3.78, 3.8, 3.8, 3.83, 3.86

c. Mode: 3.7 and 3.74   d. MR: $\frac{3.57 + 3.86}{2} = 3.715$

3.
a. $\overline{X} = \frac{\Sigma X}{n} = \frac{1566}{23} = 68.1$

b. MD: 38, 42, 42, ..., **68**, ..., 85, 90, 91

c. Mode = 42, 62, 64, 66, 72, 74   d. MR $= \frac{38 + 91}{2} = 64.5$

For the best measure of average, answers will vary.

5.
a. $\overline{X} = \frac{\Sigma X}{n} = \frac{150,755}{16} = 9422.2$

b. MD $= \frac{8632 + 9344}{2} = 8988$

c. Mode: 7552, 8632, 12568   d. MR $= \frac{6300 + 12568}{2} = 9434$

The claim seems a litle high.

7.
a. $\overline{X} = \frac{\Sigma X}{n} = \frac{79.6}{12} = 6.63$

b. MD: 5.4, 5.4, 6.2, 6.2, 6.4, **6.4, 6.5**, 7.0, 7.2, 7.2, 7.7, 8.0   MD $= \frac{6.4 + 6.5}{2} = 6.45$

c. Mode: 5.4, 6.2, 6.4, 7.2   d. MR $= \frac{5.4 + 8.0}{2} = 6.7$

For the best measure of average, answers will vary.

9.
a. $\overline{X} = \frac{\Sigma X}{n} = \frac{748.5}{16} = 46.78$

b. MD: 29.7, 32.9, 36, ..., **47.3, 48**, ...55.8, 57.2, 58.4   MD $= \frac{47.3 + 48}{2} = 47.65$

c. Mode: none   d. MR $= \frac{29.7 + 58.4}{2} = 44.05$

11.
For 2004:

a. $\overline{X} = \frac{\Sigma X}{n} = \frac{202,108}{24} = 8421.2$ b. MD $= \frac{8174 + 8220}{2} = 8197$

c. Mode: no mode   d. MR $= \frac{5044 + 14925}{2} = 9984.5$

11. continued
For 1990:

a. $\overline{X} = \frac{\Sigma X}{n} = \frac{235,440}{24} = 9810$

b. $MD = \frac{9200 + 9229}{2} = 9214.5$

Mode: no mode

d. $MR = \frac{4768 + 21923}{2} = 13,345.5$

Based on this data, it appears that the population is declining.

13.

Class Limits	Boundaries	X_m	f	$f \cdot X_m$
2.48 - 7.48	2.475 - 7.485	4.98	7	34.86
7.49 - 12.49	7.485 - 12.495	9.99	3	29.97
12.50 - 17.50	12.495 - 17.505	15.00	1	15.00
17.51 - 22.51	17.505 - 22.515	20.01	7	140.07
22.52 - 27.52	2.515 - 27.525	25.02	5	125.10
27.53 - 32.53	27.525 - 32.535	30.03	5	150.15
			28	495.15

a. $\overline{X} = \frac{\Sigma f \cdot X_m}{n} = \frac{495.15}{28} = 17.684$

b. modal class: $2.48 - 7.48$ and $17.51 - 22.51$ The grouped mean is less.

15.

Percentage	Boundaries	X_m	f	$f \cdot X_m$
0.8 - 4.4	0.75 - 4.45	2.6	26	67.6
4.5 - 8.1	4.45 - 8.15	6.3	11	69.3
8.2 - 11.8	8.15 - 11.85	10.0	4	40.0
11.9 - 15.5	11.85 - 15.55	13.7	5	68.5
15.6 - 19.2	15.55 - 19.25	17.4	2	34.8
19.3 - 22.9	19.25 - 22.95	21.1	1	21.1
23.0 - 26.6	22.95 - 26.65	24.8	1	24.8
			50	326.1

a. $\overline{X} = \frac{\Sigma f \cdot X_m}{n} = \frac{326.1}{50} = 6.52$

b. modal class: $0.8 - 4.4$

The mean is probably not the best measure of central tendency for this data because the data is "top" heavy.

17.

Percentage	Boundaries	X_m	f	$f \cdot X_m$
15.2 - 19.6	15.15 - 19.65	17.4	3	52.2
19.7 - 24.1	19.65 - 24.15	21.9	15	328.5
24.2 - 28.6	24.15 - 28.65	26.4	19	501.6
28.7 - 33.1	28.65 - 33.15	30.9	6	185.4
33.2 - 37.6	33.15 - 37.65	35.4	7	247.8
37.7 - 42.1	37.65 - 42.15	39.9	0	0
42.2 - 46.6	42.15 - 46.65	44.4	1	44.4
			51	1359.9

17. continued

a. $\overline{X} = \dfrac{\sum f \cdot X_m}{n} = \dfrac{1359.9}{51} = 26.66$ or 26.7

b. modal class: $24.2 - 28.6$

19.

Boundaries	X_m	f	$f \cdot X_m$
0.5 - 19.5	10	12	120
19.5 - 38.5	29	7	203
38.5 - 57.5	48	5	240
57.5 - 76.5	67	3	201
76.5 - 95.5	86	3	258
		30	1022

a. $\overline{X} = \dfrac{\sum f \cdot X_m}{n} = \dfrac{1022}{30} = 34.1$

b. modal class: $0.5 - 19.5$

21.

Boundaries	X_m	f	$f \cdot X_m$
15.5 - 18.5	17	14	238
18.5 - 21.5	20	12	240
21.5 - 24.5	23	18	414
24.5 - 27.5	26	10	260
27.5 - 30.5	29	15	435
30.5 - 33.5	32	6	192
		75	1779

a. $\overline{X} = \dfrac{\sum f \cdot X_m}{n} = \dfrac{1779}{75} = 23.7$

b. modal class: $21.5 - 24.5$

23.

Limits	Boundaries	X_m	f	$f \cdot X_m$
27 - 33	26.5 - 33.5	30	7	210
34 - 40	33.5 - 40.5	37	14	518
41 - 47	40.5 - 47.5	44	15	660
48 - 54	47.5 - 54.5	51	11	561
55 - 61	54.5 - 61.5	58	3	174
62 - 68	61.5 - 68.5	65	3	195
69 - 75	68.5 - 75.5	72	2	144
			55	2462

$\overline{X} = \dfrac{\sum f \cdot X_m}{n} = \dfrac{2462}{55} = 44.8$

modal class: $40.5 - 47.5$

25.

Limits	Boundaries	X_m	f	$f \cdot X_m$
1013 - 1345	1012.5 - 1345.5	1179	11	12969
1346 - 1678	1345.5 - 1678.5	1512	4	6048
1679 - 2011	1678.5 - 2011.5	1845	7	12915
2012 - 2344	2011.5 - 2344.5	2178	3	6534
2345 - 2677	2344.5 - 2677.5	2511	5	12555
2678 - 3010	2677.5 - 3010.5	2844	3	8532
			33	59553

$$\overline{X} = \frac{\sum f \cdot X_m}{n} = \frac{59553}{33} = 1804.6$$

modal class: $1013 - 1345$

27.

$$\overline{X} = \frac{\sum w \cdot X}{\sum w} = \frac{3(3.33) + 3(3.00) + 2(2.5) + 2.5(4.4) + 4(1.75)}{3 + 3 + 2 + 2.5 + 4} = \frac{41.99}{14.5} = 2.896$$

29.

$$\overline{X} = \frac{\sum w \cdot X}{\sum w} = \frac{9(427000) + 6(365000) + 12(725000)}{9 + 6 + 12} = \frac{14,733,000}{27} = \$545,666.67$$

31.

$$\overline{X} = \frac{\sum w \cdot X}{\sum w} = \frac{1(62) + 1(83) + 1(97) + 1(90) + 2(82)}{6} = \frac{496}{6} = 82.7$$

33.
a. Median   d. Mode
b. Mean   e. Mode
c. Mode   f. Mean

35.
Both could be true since one could be using the mean for the average salary, and the other could be using the mode for the average.

37.
$5 \cdot 8.2 = 41$
$6 + 10 + 7 + 12 + x = 41$
$x = 6$

39.
a. $\frac{2}{\frac{1}{30} + \frac{1}{45}} = 36$ mph

b. $\frac{2}{\frac{1}{40} + \frac{1}{25}} = 30.77$ mph

c. $\frac{2}{\frac{1}{50} + \frac{1}{10}} = \$16.67$

41.
$$\sqrt{\frac{8^2 + 6^2 + 3^2 + 5^2 + 4^2}{5}} = \sqrt{30} = 5.48$$

EXERCISE SET 3-2

1.
The square root of the variance is equal to the standard deviation.

3.
σ^2, σ

5.
When the sample size is less than 30, the formula for the true standard deviation of the sample will underestimate the population standard deviation.

7.
R = 48 − 0 = 48
$$s^2 = \frac{n\sum X^2-(\sum X)^2}{n(n-1)} = \frac{10(4061)-133^2}{10(10-1)} = \frac{22921}{90} = 254.68 \approx 254.7$$

$s = \sqrt{254.7} = 15.96 \approx 16$ The data vary widely.

9.
For Temperature:
R = 61 − 29 = 32
$$s^2 = \frac{n\sum X^2-(\sum X)^2}{n(n-1)} = \frac{10(20,777)-441^2}{10(10-1)} = 147.66 \text{ or } 147.7$$

$s = \sqrt{147.7} = 12.15$

For Precipitation:
R = 5.1 − 1.1 = 4.0
$$s^2 = \frac{n\sum X^2-(\sum X)^2}{n(n-1)} = \frac{10(86.13)-26.3^2}{10(10-1)} = 1.88$$

$s = \sqrt{1.88} = 1.37$ Temperature is more variable.

11.
Houston:
$\overline{X} = \frac{725}{13} = 55.8$
R = 75 − 47 = 28
$$s^2 = \frac{n\sum X^2-(\sum X)^2}{n(n-1)} = \frac{13(41,379)-725^2}{13(13-1)} = \frac{12,302}{13(12)} = 78.859$$

$s = \sqrt{78.859} = 8.88$

Pittsburgh:
$\overline{X} = \frac{539}{13} = 41.5$
R = 64 − 30 = 34
$$s^2 = \frac{n\sum X^2-(\sum X)^2}{n(n-1)} = \frac{13(23,413)-539^2}{13(13-1)} = \frac{13,848}{13(12)} = 88.769$$

$s = \sqrt{88.769} = 9.42$

The data for Pittsburgh seems more variable. Using the coefficient of variation to compare the data:
For Houston, $CVar = \frac{8.88}{55.8} = 15.91\%$. For Pittsburgh, $CVar = \frac{9.42}{41.5} = 22.7\%$
The data for Pittsburgh is more variable using the coefficients of variation.

13.

$R = 46 - 26 = 20$

Using the range rule of thumb, $s \approx \frac{20}{4} = 5$

15.

$R = 226 - 66 = 160$

$s^2 = \dfrac{n\sum X^2 - (\sum X)^2}{n(n-1)} = \dfrac{10(272{,}582) - 1596^2}{10(10-1)} = 1984.48 \text{ or } 1984.5$

$s = \sqrt{1984.5} = 44.5$

17.

$R = 123 - 77 = 46$

$s^2 = \dfrac{n\sum X^2 - (\sum X)^2}{n(n-1)} = \dfrac{32(328{,}430) - 3230^2}{32(32-1)} = 77.48$

$s = \sqrt{77.48} = 8.8$

19.

X_m	f	$f \cdot X_m$	$f \cdot X_m^2$
16	2	32	512
23	7	161	3703
30	12	360	10,800
37	5	185	6845
44	6	264	11,616
51	1	51	2601
58	0	0	0
65	2	130	8450
	35	1183	44527

$s^2 = \dfrac{n\sum f \cdot X_m^2 - (\sum f \cdot X_m)^2}{n(n-1)} = \dfrac{35(44{,}527) - 1183^2}{35(35-1)} = 133.58 \text{ or } 133.6$

$s = \sqrt{133.58} = 11.6$

21.

X_m	f	$f \cdot X_m$	$f \cdot X_m^2$
65	13	845	54,925
128	2	256	32,768
191	0	0	0
254	5	1270	322,580
317	1	317	100,489
380	1	380	144,400
443	0	0	0
506	1	506	256,036
569	2	1138	647,522
	25	4712	1,558,720

$s^2 = \dfrac{n\sum f \cdot X_m^2 - (\sum f \cdot X_m)^2}{n(n-1)} = \dfrac{25(1{,}558{,}720) - 4712^2}{25(25-1)} = 27{,}941.76$

$s = \sqrt{27941.76} = 167.16 \text{ or } 167.2$

23.

X_m	f	$f \cdot X_m$	$f \cdot X_m^2$
5	5	25	125
14	7	98	1372
23	10	230	5290
32	3	96	3072
41	3	123	5043
50	2	100	5000
	30	672	19,902

$$s^2 = \frac{n\sum f \cdot X_m^2 - (\sum f \cdot X_m)^2}{n(n-1)} = \frac{30(19,902) - 672^2}{30(30-1)} = 167.2$$

$$s = \sqrt{167.2} = 12.93$$

25.

X_m	f	$f \cdot X_m$	$f \cdot X_m^2$
68	5	340	23,120
79	14	1106	87,374
90	18	1620	145,800
101	25	2525	255,025
112	12	1344	150,528
123	6	738	90,774
	80	7673	752,621

$$s^2 = \frac{n\sum f \cdot X_m^2 - (\sum f \cdot X_m)^2}{n(n-1)} = \frac{80(752,621) - 7673^2}{80(80-1)} = 211.19 \text{ or } 211.2$$

$$s = \sqrt{211.2} = 14.5$$

No, the variability of the lifetimes of the batteries is quite large.

27.

X_m	f	$f \cdot X_m$	$f \cdot X_m^2$
27	5	135	3645
30	9	270	8100
33	32	1056	34848
36	30	720	25920
39	12	468	18252
62	2	84	3528
	80	2733	94293

$$s^2 = \frac{n\sum f \cdot X_m^2 - (\sum f \cdot X_m)^2}{n(n-1)} = \frac{80(94,293) - 2733^2}{80(80-1)} = 11.7$$

$$s = \sqrt{11.7} = 3.4$$

29.
For US: $\overline{X} = 3386.6$, s = 693.9; C. Var = $\frac{s}{\overline{X}} = \frac{693.9}{3386.6} = 0.2049$ or 20.49%

For World: $\overline{X} = 4997.8$, s = 803.2; C. Var = $\frac{s}{\overline{X}} = \frac{803.2}{4997.8} = 0.1607$ or 16.07%

The data for US is more variable.

31.
C. Var $= \frac{s}{\bar{X}} = \frac{6}{26} = 0.231 = 23.1\%$
C. Var $= \frac{s}{\bar{X}} = \frac{4000}{31,000} = 0.129 = 12.9\%$
Age is more variable.

33.
a. $1 - \frac{1}{5^2} = 0.96$ or 96%

b. $1 - \frac{1}{4^2} = 0.9375$ or 93.75%

35.
$\bar{X} = 240$ and $s = 38$
At least 75% of the data values will fall withing two standard deviations of the mean; hence,
$2(38) = 76$ and $240 - 76 = 164$ and $240 + 76 = 316$. Hence at least 75% of the
data values will fall between 164 and 316 calories.

37.
$1 - \frac{1}{k^2} = 0.8889$
$k = 3$
$\bar{X} = 640$ and $s = 85$
At least 88.89% of the data values will fall within 3 standard deviations of the mean, hence
$640 - 3(85) = 385$ and $640 + 3(85) = 895$. Therefore at least 88.89% of the data values will fall between
385 and 895 pounds.

39.
$\bar{X} = 12$ and $s = 3$
$20 - 12 = 8$ and $8 \div 3 = 2.67$
Hence, $1 - \frac{1}{k^2} = 1 - \frac{1}{2.67^2} = 1 - 0.14 = 0.86 = 86\%$
At least 86% of the data values will fall between 4 and 20.

41.
$26.8 + 1(4.2) = 31$
By the Empirical Rule, 68% of consumption is within 1 standard deviation of the mean. Then $\frac{1}{2}$ of 32%, or
16%, of consumption would be more than 31 pounds of citrus fruit per year.

43.
$n = 30$ $\bar{X} = 214.97$ $s = 20.76$ At least 75% of the data values will fall between $\bar{X} \pm 2s$.
$\bar{X} - 2(20.76) = 214.97 - 41.52 = 173.45$ and $\bar{X} + 2(20.76) = 214.97 + 41.52 = 256.49$
In this case all 30 values fall within this range; hence Chebyshev's Theorem is correct for
this example.

45.
For $k = 1.5$, $1 - \frac{1}{1.5^2} = 1 - 0.44 = 0.56$ or 56%
For $k = 2$, $1 - \frac{1}{2^2} = 1 - 0.25 = 0.75$ or 75%
For $k = 2.5$, $1 - \frac{1}{2.5^2} = 1 - 0.16 = 0.84$ or 84%
For $k = 3$, $1 - \frac{1}{3^2} = 1 - 0.1111 = .8889$ or 88.89%
For $k = 3.5$, $1 - \frac{1}{3.5^2} = 1 - 0.08 = 0.92$ or 92%

47.
$\bar{X} = 13.3$
Mean Dev $= \frac{|5-13.3|+|9-13.3|+|10-13.3|+|11-13.3|+|11-13.3|}{10}$

$+ \frac{|12-13.3|+|15-13.3|+|18-13.3|+|20-13.3|+|22-13.3|}{10} = 4.36$

49.

For $n = 25$, $\overline{X} = 50$, and $s = 3$:

$$s\sqrt{n-1} = 3\sqrt{25-1} = 14.7 \qquad \overline{X} + s\sqrt{n-1} = 50 + 14.7 = 64.7$$

67 must be an incorrect data value, since it is beyond the range using the formula $s\sqrt{n-1}$.

EXERCISE SET 3-3

1.
A z score tells how many standard deviations the data value is above or below the mean.

3.
A percentile is a relative measure while a percent is an absolute measure of the part to the total.

5.
$Q_1 = P_{25}$, $Q_2 = P_{50}$, $Q_3 = P_{75}$

7.
$D_1 = P_{10}$, $D_2 = P_{20}$, $D_3 = P_{30}$, etc

9.
For Canada:
$z = \frac{X - \overline{X}}{s} = \frac{26 - 29.4}{8.6} = -0.40$
For Italy:
$z = \frac{X - \overline{X}}{s} = \frac{42 - 29.4}{8.6} = 1.47$
For US:
$z = \frac{X - \overline{X}}{s} = \frac{13 - 29.4}{8.6} = -1.91$

11.
a. $z = \frac{X - \overline{X}}{s} = \frac{79 - 76}{5} = 0.6$

b. $z = \frac{70 - 76}{5} = -1.2$

c. $z = \frac{88 - 76}{5} = 2.4$

d. $z = \frac{65 - 76}{5} = -2.2$

e. $z = \frac{77 - 76}{5} = 0.2$

13.
For the statistics test: $z = \frac{75 - 60}{10} = 1.5$

For the accounting test: $z = \frac{36 - 30}{\sqrt{16}} = 1.5$

Neither. The scores have the same relative position.

15.
a. $z = \frac{3.2 - 4.6}{1.5} = -0.93$ b. $z = \frac{630 - 800}{200} = -0.85$ c. $z = \frac{43 - 50}{5} = -1.4$

The score in part b is the highest.

17.
a. 24th b. 67th c. 48th d. 88th

19.
a. a. 234 b. 251 c. 263 d. 274 e. 284

21.
a. 13th b. 40th c. 54th d. 76th e. 92nd

23.
$c = \frac{9(40)}{100} = 3.6$ or 4th data value, which is 597

25.
$c = \frac{n \cdot p}{100} = \frac{7(60)}{100} = 4.2$ or 5   Hence, 47 is the closest value to the 60th percentile.

27.
$c = \frac{10(40)}{100} = 4$ average the 4th and 5th values: $P_{40} = \frac{2.1 + 2.2}{2} = 2.15$

29.
$c = \frac{6(33)}{100} = 1.98$ or 2nd data value, which is 12.

31.
a. 5, 12, 16, 25, 32, 38 $Q_1 = 12$, $Q_2 = 20.5$, $Q_3 = 32$
Midquartile $= \frac{12 + 32}{2} = 22$ Interquartile range: $32 - 12 = 20$

b. 53, 62, 78, 94, 96, 99, 103 $Q_1 = 62$, $Q_2 = 94$, $Q_3 = 99$
Midquartile $= \frac{62 + 99}{2} = 80.5$ Interquartile range: $99 - 62 = 37$

EXERCISE SET 3-4

1. Data arranged in order: 6, 8, 12, 19, 27, 32, 54
Minimum: 6
Q_1: 8
Median: 19
Q_3: 32
Maximum: 54
Interquartile Range: $32 - 8 = 24$

3. Data arranged in order: 188, 192, 316, 362, 437, 589
Minimum: 188
Q_1: 192
Median: $\frac{316 + 362}{2} = 339$
Q_3: 437
Maximum: 589
Interquartile Range: $437 - 192 = 245$

5. Data arranged in order: 14.6, 15.5, 16.3, 18.2, 19.8
Minimum: 14.6
Q_1: $\frac{14.6 + 15.5}{2} = 15.05$
Median: 16.3
Q_3: $\frac{18.2 + 19.8}{2} = 19.0$
Maximum: 19.8
Interquartile Range: $19.0 - 15.05 = 3.95$

7. Minimum: 3
Q_1: 5
Median: 8
Q_3: 9
Maximum: 11
Interquartile Range: $9 - 5 = 4$

9. Minimum: 55
Q_1: 65
Median: 70
Q_3: 90
Maximum: 95
Interquartile Range: $90 - 65 = 25$

11.
$MD = \frac{30 + 31}{2} = 30.5$
$Q_1 = 29 \quad Q_3 = 34$

```
        27      29   30.5           34
              •---------┌────┬──────────┐
                        │    │          │
                        └────┴──────────┘
  ─────────────────────────────────────────
  |       |       |       |       |       |
  25      27      29      31      33      35
```

13.
$MD = 10$
$Q_1 = 8 \qquad Q_3 = 16$

```
 5     8   10       16      20
 •------┌───┬────────┐---------•
        │   │        │
        └───┴────────┘
  ──────────────────────────
  |        |        |       |
  5        10       15      20
```

The box plot of the data is somewhat positively skewed.

15.
$MD = \frac{123 + 123}{2} = 123$
$Q_1 = 29.5 \qquad Q_3 = 135.5$

```
 10  29.5            123 135.5                       316
  •--┌─────────────────┬──┐-----------------------------•
     │                 │  │
     └─────────────────┴──┘
  ───────────────────────────────────────────────
  |      |      |      |      |      |      |      |
  0      50     100    150    200    250    300    350
```

Based on the position of the median, the distribution is negatively skewed. Based on the lengths of the lines, the distribution is positively skewed.

17.
(a)
For April: $\overline{X} = 138$
For May: $\overline{X} = 391.7$
For June: $\overline{X} = 292$

17a. continued
For July: $\overline{X} = 143$
The month with the highest mean number of tornadoes is May.

(b)
For 2005: $\overline{X} = 177.3$
For 2004: $\overline{X} = 256.5$
For 2003: $\overline{X} = 289.8$
The year with the highest mean number of tornadoes is 2003.

(c) The 5-number summaries for each year are:
For 2005: 123, 127.5, 135, 227, 316
For 2004: 124, 124.5, 196.5, 388.5, 509
For 2003: 157, 162, 229.5, 417.5, 543

```
157 162  229.5              417.5        543
     ┌─────┬─────────────┐
 ●───┤     │             ├------------●              For 2003


124 124.5 196.5      388.5        509
     ┌─────┬──────────┐
 ●───┤     │          ├-------------●                For 2004


123 127.5 135 227    316
   ┌──┬────┐
 ●─┤  │    ├--------●                                For 2005


  │      │      │      │      │      │
 100    200    300    400    500    600
```

The distribution for 2003, 2004, and 2005 are positively skewed. The data for 2005 appears to be the least variable.

REVIEW EXERCISES - CHAPTER 3

1.
a. $\overline{X} = \dfrac{\sum X}{n} = \dfrac{272}{10} = 27.2$

b. 17, 17, 17, 18, **19, 19**, 26, 28, 52, 59

$MD = \dfrac{19+19}{2} = 19$

c. Mode $= 17$

d. $MR = \dfrac{17+59}{2} = 38$

e. Range $= 59 - 17 = 42$

f. $s^2 = \dfrac{n\sum X^2 - (\sum X)^2}{n(n-1)} = \dfrac{10(9558) - 272^2}{10(10-1)} = 239.96$

g. $s = \sqrt{239.96} = 15.5$

3.

Class	X_m	f	$f \cdot X_m$	$f \cdot X_m^2$	cf
1 - 3	2	1	2	4	1
4 - 6	5	4	20	100	5
7 - 9	8	5	40	320	10
10 - 12	11	1	11	121	11
13 - 15	14	1	14	196	12
		12	87	741	

a. $\overline{X} = \dfrac{\sum f \cdot X_m}{n} = \dfrac{87}{12} = 7.3$

b. Modal Class $= 7 - 9$ or $6.5 - 9.5$

c. $s^2 = \dfrac{12(741) - 87^2}{12(11)} = \dfrac{1323}{132} = 10.0$

d. $s = \sqrt{10.0} = 3.2$

5.

Class Boundaries	X_m	f	$f \cdot X_m$	$f \cdot X_m^2$	cf
12.5 - 27.5	20	6	120	2400	6
27.5 - 42.5	35	3	105	3675	9
42.5 - 57.5	50	5	250	12,500	14
57.5 - 72.5	65	8	520	33,800	22
72.5 - 87.5	80	6	480	38,400	28
87.5 - 102.5	95	2	190	18,050	30
		30	1665	108,825	

a. $\overline{X} = \dfrac{\sum f \cdot X_m}{n} = \dfrac{1665}{30} = 55.5$

b. Modal class $= 57.5 - 72.5$

c. $s^2 = \dfrac{n\sum f \cdot X_m^2 - (\sum f \cdot X_m)^2}{n(n-1)} = \dfrac{30(108,825) - 1665^2}{30(30-1)} = 566.1$

d. $s = \sqrt{566.1} = 23.8$

7.

$\overline{X} = \dfrac{\sum w \cdot X}{\sum w} = \dfrac{1.6(1.4) + 0.8(0.8) + 0.4(0.3) + 1.8(1.6)}{1.4 + 0.8 + 0.3 + 1.6} = 1.43$ viewers per household

9.

$\overline{X} = \dfrac{\sum w \cdot X}{\sum w} = \dfrac{8 \cdot 3 + 1 \cdot 6 + 1 \cdot 30}{8 + 1 + 1} = \dfrac{60}{10} = 6$

11.

Magazines: C. Var $= \dfrac{s}{\overline{X}} = \dfrac{12}{56} = 0.214$ or 21.4%

Year: C. Var $= \dfrac{s}{\overline{X}} = \dfrac{2.5}{6} = 0.417$ or 41.7%

The number of years is more variable.

13.
a.

b. $P_{35} = 50$; $P_{65} = 53$; $P_{85} = 55$ (answers are approximate)
c. $44 = 10^{th}$ percentile; $48 = 26^{th}$ percentile; $54 = 78^{th}$ percentile
(answers are approximate)

15.
$\overline{X} = 0.32$ $s = 0.03$ $k = 2$
$0.32 - 2(0.03) = 0.26$ and $0.32 + 2(0.03) = 0.38$
At least 75% of the values will fall between \$0.26 and \$0.38.

17.
$\overline{X} = 54$ $s = 4$ $60 - 54 = 6$ $k = \frac{6}{4} = 1.5$ $1 - \frac{1}{1.5^2} = 1 - 0.44 = 0.56$ or 56%

19.
$\overline{X} = 32$ $s = 4$ $44 - 32 = 12$ $k = \frac{12}{4} = 3$ $1 - \frac{1}{3^2} = 0.8889 = 88.89\%$

21.
Top Nine Movies:
$MD = 2820$ $Q_1 = 2520.5$ $Q_3 = 3127.5$

Top Ten Movies:
$MD = 2699.5$ $Q_1 = 2516$ $Q_3 = 3044$

```
                    2820
   2330   2520    3127.5    3687
    •-----[    |    ]--------•            Top Nine
   _____
    |        |        |        |        |
  2000     2500     3000     3500     4000
```

The range is much larger.

23. By the Empirical Rule, 68% of the scores will be within 1 standard deviation of the mean.
$29.7 + 1(6) = 35.7$
$29.7 - 1(6) = 23.7$
Then 68% of commuters will get to work between 23.7 and 35.7 minutes.

CHAPTER 3 QUIZ

1. True
2. True

3. False
4. False
5. False
6. False
7. False
8. False
9. False
10. c
11. c
12. a and b
13. b
14. d
15. b
16. Statistic
17. Parameters, statistics
18. Standard deviation
19. σ
20. Midrange
21. Positively
22. Outlier
23. a. 15.3   b. 15.5   c. 15, 16, 17   d. 15   e. 6   f. 3.57   g. 1.9
24. a. 6.4   b. 6 − 8   c. 11.6   d. 3.4
25. a. 51.4   b. 35.5 − 50.5   c. 451.5   d. 21.2
26. a. 8.2   b. 7 − 9   c. 21.6   d. 4.6
27. 1.6
28. 4.46 or 4.5
29. 0.33; 0.162; newspapers
30. 0.3125; 0.229; brands
31. − 0.75; − 1.67; science
32. a. 0.5   b. 1.6   c. 15, c is higher
33. a. 56.25; 43.75; 81.25; 31.25; 93.75; 18.75; 6.25; 68.75   b. 0.9
    c.

34.
a.

b. 47; 55; 64

34. continued
c. 56th percentile; 6th percentile; 99th percentile

35.
For Pre-buy:
MD = 1.625 or 1.63 $Q_1 = 1.54$ $Q_3 = 1.65$
For No Pre-buy:
MD = 3.95 $Q_1 = 3.85$ $Q_3 = 3.99$

```
        1.45   1.54 1.63 1.65  1.72
         •--------┌─────┬───┐------•
                  │     │   │
    _____
         |       |      |       |       |
        1.40    1.50   1.60    1.70    1.80
                Pre-buy Cost
```

```
     3.80  3.85     3.95 3.99           4.19
       •---┌──────┬──────┐-------------------•
           │      │      │
    _____
         |       |      |       |       |
        3.80    3.90   4.00    4.10    4.20
                No Pre-buy Cost
```

The cost of pre-buy gas is much less than to return the car without filling it with gas. The variability of the return without filling with gas is larger than the variability of the pre-buy gas.

36.
For above 1129: 16%
For above 799: 97.5%

Note: Answers may vary due to rounding, TI-83's or computer programs.

EXERCISE SET 4-1

1.
A probability experiment is a chance process which leads to well-defined outcomes.

3.
An outcome is the result of a single trial of a probability experiment, whereas an event can consist of one or more outcomes.

5.
The range of values is $0 \leq P(E) \leq 1$.

7.
0

9.
$1 - 0.20 = 0.80$
Since the probability that it won't rain is 80%, you could leave your umbrella at home and be fairly safe.

11.
a. empirical
b. classical
c. empirical
d. classical
e. empirical
f. empirical
g. subjective

13.
There are 6^2 or 36 outcomes.
a. There are 4 ways to get a sum of 9. They are (6,3), (5,4), (4,5), and (3,6). The probability then is $\frac{4}{36} = \frac{1}{9}$.

b. There are 6 ways to get a sum of 7 and 2 ways to get a sum of 11. They are (6,1), (5,2), (4,3), (3,4), (2,5), (1,6), (6,5) and (5,6). The probability then is $\frac{8}{36} = \frac{2}{9}$.

c. There are 6 ways to get a doubles. They are (1,1), (2,2), (3,3), (4,4), (5,5), and (6,6). The probability then is $\frac{6}{36} = \frac{1}{6}$.

d. To get a sum less than nine, one must roll a 2, 3, 4, 5, 6, 7, or 8. There are 26 ways to get a sum less than 9. The probability then is $\frac{26}{36} = \frac{13}{18}$.

e. To get a sum greater than or equal to 10, one must roll a 10, 11, or 12. There are six

13e. continued
ways to do this. They are (6,4), (5,5), (4,6), (6,5), (5,6), and (6,6). The probability is $\frac{6}{36} = \frac{1}{6}$.

14.
a. $\frac{1}{13}$ f. $\frac{1}{52}$
b. $\frac{1}{4}$ g. $\frac{1}{26}$
c. $\frac{1}{52}$ h. $\frac{1}{26}$
d. $\frac{2}{13}$ i. $\frac{1}{2}$
e. $\frac{4}{13}$ j. $\frac{1}{2}$

15.
There are 20 possible outcomes.

a. P(winning $10) = P(rolling a 1)
P(rolling a 1) = $\frac{2}{20} = \frac{1}{10} = 0.1$

b. P(winning $5 or $10) = P(rolling either a 1 or 2)
P(1 or 1) = $\frac{4}{20} = \frac{1}{5} = 0.2$

c. P(winning a coupon) = P(rolling either a 3 or 4)
P(3 or 4) = $\frac{16}{20} = \frac{4}{5} = 0.8$

17.
a. P(type O) = 0.43

b. P(type A or B) = $0.40 + 0.12 = 0.52$

c. P(not type A or O) = $1 - 0.83 = 0.17$

19.
a. P(even prime number) = $\frac{1}{25} = 0.04$

b. P(sum of the digits is even) = $\frac{13}{25} = 0.52$

c. P(greater than 50) = $\frac{10}{25} = 0.4$

21.
The sample space is BBB, BBG, BGB, GBB, GGB, GBG, BGG, and GGG.

a. All boys is the outcome BBB; hence P(all boys) = $\frac{1}{8}$.

b. All girls or all boys would be BBB and GGG; hence, P(all girls or all boys) = $\frac{1}{4}$.

21. continued

c. Exactly two boys or two girls would be BBG, BGB, GBB, BBG, GBG, or BGG. The probability then is $\frac{6}{8} = \frac{3}{4}$.

d. At least one child of each gender means at least one boy or at least one girl. The outcomes are the same as those of part c, hence the probability is the same, $\frac{3}{4}$.

23.

The outcomes for 2, 3, or 12 are (1,1), (1,2), (2,1), and (6,6); hence P(2, 3, or 12) = $\frac{1+2+1}{36} = \frac{4}{36} = \frac{1}{9}$.

25.

a. P(debt is less than $5001) = 27%.

b. P(debt is more than $20,000) = P($20,001 to $50,000) + P($50,000+) = 19% + 14% = 33%

c. P(debt is between $1 and $20,000) = P($1 to $5000) + P($5001 to $20,000) = 27% + 40% = 67%

d. P(debt is more than $50,000) = 14%

27.

P(ten thousand dollar bill) = $\frac{3,460,000}{5,225,000} = 0.66$

29.

(a)

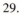

	1	2	3	4	5	6
1	1	2	3	4	5	6
2	2	4	6	8	10	12
3	3	6	9	12	15	18
4	4	8	12	16	20	24
5	5	10	15	20	25	30
6	6	12	18	24	30	36

(b) P(multiple of 6) = $\frac{15}{36} = \frac{5}{12}$

(c) P(less than 10) = $\frac{17}{36}$

31.

33.

35.

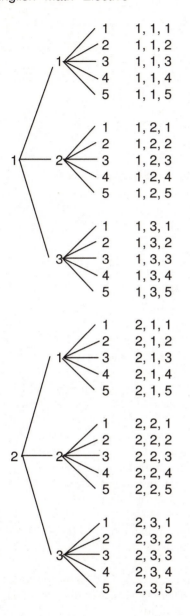

37.
a. 0.08
b. 0.01
c. $0.08 + 0.27 = 0.35$
d. $0.01 + 0.24 + 0.11 = 0.36$

39.
The statement is probably not based on empirical probability and probably not true.

41.
Actual outcomes will vary, however each number should occur approximately $\frac{1}{6}$ of the time.

43.
a. 1:5, 5:1   e. 1:12, 12:1
b. 1:1, 1:1   f. 1:3, 3:1
c. 1:3, 3:1   g. 1:1, 1:1
d. 1:1, 1:1

EXERCISE SET 4-2

1.
Two events are mutually exclusive if they cannot occur at the same time. Examples will vary.

3.
a. $\frac{1,348,503}{1,907,172} = 0.707$

b. $\frac{46,024}{1,907,172} + \frac{1,098,371}{1,907,172} - \frac{21,683}{1,907,172} =$

$\frac{1,122,712}{1,907,172} = 0.589$

c. $\frac{21,683}{1,907,172} = 0.011$

d. $\frac{1,394,527}{1,907,172} = 0.731$

5.
$\frac{4}{19} + \frac{7}{19} = \frac{11}{19}$

7.
a. $\frac{3}{22} + \frac{7}{22} + \frac{4}{22} = \frac{14}{22}$ or $\frac{7}{11}$
b. $\frac{5}{22} + \frac{7}{22} = \frac{12}{22}$ or $\frac{6}{11}$
c. $\frac{3}{22} + \frac{7}{22} + \frac{4}{22} = \frac{14}{22}$ or $\frac{7}{11}$
d. $\frac{7}{22} + \frac{4}{22} = \frac{11}{22}$ or $\frac{1}{2}$

9.
P(football or basketball) =
$\frac{58 + 40 - 8}{200} = \frac{90}{200}$ or 0.45

P(neither) $= 1 - \frac{90}{200} = \frac{11}{20}$ or 0.55

11.

	Junior	Senior	Total
Female	6	6	12
Male	12	4	16
Total	18	10	28

a. $\frac{18}{28} + \frac{12}{28} - \frac{6}{28} = \frac{24}{28} = \frac{6}{7}$

11. continued

b. $\frac{10}{28} + \frac{12}{28} - \frac{6}{28} = \frac{16}{28} = \frac{4}{7}$

c. $\frac{18}{28} + \frac{10}{28} = \frac{28}{28} = 1$

13.

	18 - 24	25 - 34	Total
Male	7922	2534	10,456
Female	5779	995	6,774
Total	13,701	3529	17,230

a. P(female aged 25 - 34) $= \frac{995}{17,230} = 0.058$

b. P(male or aged 18 - 24) =

$\frac{10,456}{17,230} + \frac{13,701}{17,230} - \frac{7922}{17,230} =$

$\frac{16,235}{17,230} = 0.942$

c. P(under 25 years and not male) =

$\frac{5779}{17,230} = 0.335$

15.

Total = 136,238 multiple births

a. P(more than two babies) =
$\frac{7663}{136,328} = 0.056$

b. P(quads or quints) $= \frac{553}{136,328} = 0.004$

c. The total number of babies who are triplets = 21,330

The total number of babies from multiple births = 280,957

P(baby is a triplet) $= \frac{21,330}{280,957} = 0.076$

17.

Class	Yes	No	No Opinion	Total
Freshmen	15	8	5	28
Sophomores	24	4	2	30
Total	39	12	7	58

a. P(no opinion) $= \frac{7}{58}$

b. P(is a freshman or is against the issue) =
P(freshman) + P(against) − P(both) =
$\frac{28}{58} + \frac{12}{58} - \frac{8}{58} = \frac{32}{58}$ or $\frac{16}{29}$

c. P(sophomore and in favor) $= \frac{24}{58}$ or $\frac{12}{29}$

19.

The total of the frequencies is 30.

19. continued

a. $\frac{2}{30} = \frac{1}{15}$

b. $\frac{2+3+5}{30} = \frac{10}{30} = \frac{1}{3}$

c. $\frac{12+8+2+3}{30} = \frac{25}{30} = \frac{5}{6}$

d. $\frac{12+8+2+3}{30} = \frac{25}{30} = \frac{5}{6}$

e. $\frac{8+2}{30} = \frac{10}{30} = \frac{1}{3}$

21.

The total of the frequencies is 24.

a. $\frac{10}{24} = \frac{5}{12}$

b. $\frac{2+1}{24} = \frac{3}{24} = \frac{1}{8}$

c. $\frac{10+3+2+1}{24} = \frac{16}{24} = \frac{2}{3}$

d. $\frac{8+10+3+2}{24} = \frac{23}{24}$

23.

a. There are 4 kings, 4 queens, and 4 jacks; hence P(king or queen or jack) $= \frac{12}{52} = \frac{3}{13}$

b. There are 13 clubs, 13 hearts, and 13 spades; hence, P(club or heart or spade) = $\frac{13+13+13}{52} = \frac{39}{52} = \frac{3}{4}$

c. There are 4 kings, 4 queens, and 13 diamonds but the king and queen of diamonds were counted twice.
Hence; P(king or queen or diamond) = P(king) + P(queen) + P(diamond) − P(king and queen of diamonds) $= \frac{4}{52} + \frac{4}{52} + \frac{13}{52} - \frac{2}{52} = \frac{19}{52}$

d. There are 4 aces, 13 diamonds, and 13 hearts. There is one ace of diamonds and one ace of hearts.
Hence, P(ace or diamond or heart) = P(ace) + P(diamond) + P(heart) − P(ace of hearts and ace of diamonds) = $\frac{4}{52} + \frac{13}{52} + \frac{13}{52} - \frac{2}{52} = \frac{28}{52} = \frac{7}{13}$

e. There are 4 nines, 4 tens, 13 spades, and 13 clubs. There is one nine of spades, one ten of spades, one nine of clubs and one ten of clubs.
Hence, P(9 or 10 or spade or club) = P(9) + P(10) + P(spade) + P(club) − P(9 and 10 of clubs and spades) = $\frac{4}{52} + \frac{4}{52} + \frac{13}{52} + \frac{13}{52} - \frac{4}{52} = \frac{30}{52} = \frac{15}{26}$

25.
P(export or
ethanol) $= \frac{1.9}{11} + \frac{1.6}{11} - \frac{0}{11} = 0.318$

27.
P(mushrooms or pepperoni) =
    P(mushrooms) + P(pepperoni) −
    P(mushrooms and pepperoni)

Let X = P(mushrooms and pepperoni)
Then $0.55 = 0.32 + 0.17 - X$
$X = 0.06$

29.
P(not a two-car garage) $= 1 - 0.70 = 0.30$

EXERCISE SET 4-3

1.
a. independent   e. independent
b. dependent   f. dependent
c. dependent   g. dependent
d. dependent   h. independent

3.
a. P(none play video or computer
games) $= (0.31)^4 = 0.009$ or 0.9%

b. P(all four play video or computer
games) $= (0.69)^4 = 0.227$ or 22.7%

5.
P(making a sale) $= 0.21$
P(making 4 sales) $= (0.21)^4 = 0.00194$
The event is unlikely to occur since the
probability is small.

7.
P(all are citizens) $= (0.801)^3 = 0.5139$

9.
P(none use phones for
texting) $= (0.65)^4 = 0.179$

11.
a. P(none had cable TV) $= (0.14)^3 = 0.003$

b. P(all three had cable
TV) $= (0.86)^3 = 0.636$

c. P(at least one had cable TV)
$= 1 - 0.003$
$= 0.997$

13.
a. $\frac{4}{52} \cdot \frac{3}{51} \cdot \frac{2}{50} \cdot \frac{1}{49} = \frac{1}{270,725}$

b. $\frac{13}{52} \cdot \frac{12}{51} \cdot \frac{11}{50} \cdot \frac{10}{49} = \frac{11}{4165}$

c. $\frac{26}{52} \cdot \frac{25}{51} \cdot \frac{24}{50} \cdot \frac{23}{49} = \frac{46}{833}$

15.
$\frac{3}{8} \cdot \frac{2}{7} \cdot \frac{1}{6} = \frac{1}{56}$

17.
P(both prizes are won by women) =
$\frac{20}{30} \cdot \frac{19}{29} = \frac{38}{87}$ more likely than in #16.

19.

a. P(male student without
aid) $= 0.4237(0.394) = 0.0167$

b. P(male student | student has aid) =

$\frac{\text{P(aid \& male)}}{\text{P(aid)}} = \frac{0.4237(0.606)}{0.4237(0.606)+0.5763(0.652)} = 0.406$

c. P(female student or a student who
receives federal aid) =
P(female) + P(federal aid) − P(female with
federal aid) =
$0.5763 + (0.115 + 0.1894) - 0.1894 = 0.69$

21.

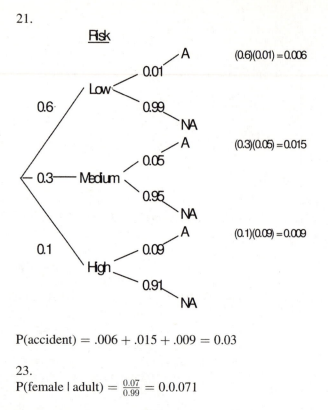

Risk

$P(\text{accident}) = .006 + .015 + .009 = 0.03$

23.
$P(\text{female} \mid \text{adult}) = \frac{0.07}{0.99} = 0.0.071$

25.
$P(\text{2nd defective} \mid \text{1st defective}) =$

$\frac{P(\text{1st and 2nd defective})}{P(\text{1st defective})} = \frac{\frac{2}{8} \cdot \frac{1}{7}}{\frac{2}{8}} = \frac{1}{7}$

27.
$P(\text{calculus} \mid \text{dean's list}) = \frac{0.042}{0.21} = 0.2$

29.
$P(\text{salad} \mid \text{pizza}) = \frac{0.65}{0.95} = 0.684$ or 68.4%

31.
a. $P(O^-) = 0.06$
b. $P(\text{type O} \mid Rh^+) = \frac{0.37}{0.85} = 0.4353$
c. $P(A^+ \text{ or } AB^-) = 0.34 + 0.01 = 0.35$
d. $P(Rh^- \mid \text{type B}) = \frac{0.02}{0.12} = 0.1667$

33.
a. $P(\text{gold} \mid \text{U. S.}) = \frac{\frac{36}{957}}{\frac{110}{957}} = 0.327$

b. $P(\text{U. S.} \mid \text{gold}) = \frac{\frac{36}{957}}{\frac{302}{957}} = 0.119$

c. No. $P(\text{gold} \mid \text{U. S.}) \neq P(\text{gold})$

35.
a. $P(\text{all 3 get enough}$
exercise$) = (0.27)^3 = 0.0197$

35. continued
b. $P(\text{at least one gets enough}$
exercise$) = 1 - (0.73)^3 = 0.611$

37.
a. $P(\text{none have been}$
married$) = (0.703)^5 = 0.1717$

b. $P(\text{at least one has been married}) =$
$1 - P(\text{none have been married})$
$= 1 - 0.1717$
$= 0.8283$

39.
$P(\text{at least one not on time}) =$
$1 - P(\text{none not on time})$
$= 1 - P(\text{all 5 on time})$
$= 1 - (0.843)^5 = 0.574$

41.
If $P(\text{read to}) = 0.58$, then
$P(\text{not being read to}) = 1 - 0.58 = 0.42$

$P(\text{at least one is read to}) = 1 - P(\text{none are}$
read to$)$
$= 1 - P(\text{all five are not read to})$
$= 1 - (0.42)^5 = 0.9869$

43.
$P(\text{at least one club}) = 1 - P(\text{no clubs})$
$1 - \frac{39}{52} \cdot \frac{38}{51} \cdot \frac{37}{50} \cdot \frac{36}{49} = 1 - \frac{6327}{20,825}$
$= \frac{14,498}{20,825}$

45.
a. $P(\text{not a family and children's}$
game$) = 1 - 0.198 = 0.802$
$P(\text{none of five are family and children's}$
games$) = (0.802)^5 = 0.332$

b. $P(\text{at least one is family and children's}$
game$) = 1 - 0.332 = 0.668$

47.
$P(\text{at least one tail}) = 1 - P(\text{no tails})$
$1 - \left(\frac{1}{2}\right)^5 = 1 - \frac{1}{32} = \frac{31}{32}$

49.
$P(\text{rolling a 4}) = \frac{1}{6}$
$P(\text{at least one 4}) = 1 - P(\text{no fours})$
$P(\text{at least one 4}) = 1 - P(\text{all 6 are not 4's})$
$P(\text{at least one 4}) = 1 - \left(\frac{5}{6}\right)^6 = 0.665$
It will happen almost 67% $\left(\frac{2}{3}\right)$ of the time.
It's somewhat likely.

51.

P(at least one even) = 1 − P(no evens)

$1 - (\frac{1}{2})^3 = 1 - \frac{1}{8} = \frac{7}{8}$

53.

No, because $P(A \cap B) = 0$ and
$P(A \cap B) \neq P(A) \cdot P(B)$

55.

Yes.

P(enroll) = 0.55

P(enroll | DW) > P(enroll) which indicates that DW has a positive effect on enrollment.

P(enroll | LP) = P(enroll) which indicates that LP has no effect on enrollment.

P(enroll | MH) < P(enroll) which indicates that MH has a low effect on enrollment.

Thus, all students should meet with DW.

EXERCISE SET 4-4

1.

$10^5 = 100,000$

$10 \cdot 9 \cdot 8 \cdot 7 \cdot 6 = 30,240$

3.

$6! = 6 \cdot 5 \cdot 4 \cdot 3 \cdot 2 \cdot 1 = 720$

5.

$7! = 7 \cdot 6 \cdot 5 \cdot 4 \cdot 3 \cdot 2 \cdot 1 = 5040$

7.

$10! = 7 \cdot 6 \cdot 5 \cdot 4 \cdot 3 \cdot 2 \cdot 1 = 5040$

9.

$10 \cdot 10 \cdot 10 = 1000$

$1 \cdot 9 \cdot 8 = 72$

11.

$6 \cdot 5 \cdot 5 \cdot 4 = 600$

13.

a. $8! = 8 \cdot 7 \cdot 6 \cdot 5 \cdot 4 \cdot 3 \cdot 2 \cdot 1 = 40,320$

b. $10! = 10 \cdot 9 \cdot 8 \cdot 7 \cdot 6 \cdot 5 \cdot 4 \cdot 3 \cdot 2 \cdot 1$
$10! = 3,628,800$

c. $0! = 1$

13. continued

d. $1! = 1$

e. $_7P_5 = \frac{7!}{(7-5)!}$

$= \frac{7 \cdot 6 \cdot 5 \cdot 4 \cdot 3 \cdot 2 \cdot 1}{2 \cdot 1} = 2520$

f. $_{12}P_4 = \frac{12!}{(12-4)!}$

$= \frac{12 \cdot 11 \cdot 10 \cdot 9 \cdot 8 \cdot 7 \cdot 6 \cdot 5 \cdot 4 \cdot 3 \cdot 2 \cdot 1}{8 \cdot 7 \cdot 6 \cdot 5 \cdot 4 \cdot 3 \cdot 2 \cdot 1} = 11,880$

g. $_5P_3 = \frac{5!}{(5-3)!}$

$= \frac{5 \cdot 4 \cdot 3 \cdot 2 \cdot 1}{2 \cdot 1} = 60$

h. $_6P_0 = \frac{6!}{(6-0)!}$

$= \frac{6 \cdot 5 \cdot 4 \cdot 3 \cdot 2 \cdot 1}{6 \cdot 5 \cdot 4 \cdot 3 \cdot 2 \cdot 1} = 1$

i. $_5P_5 = \frac{5!}{(5-5)!}$

$= \frac{5 \cdot 4 \cdot 3 \cdot 2 \cdot 1}{0!} = 120$

j. $_6P_2 = \frac{6!}{(6-2)!}$

$= \frac{6 \cdot 5 \cdot 4 \cdot 3 \cdot 2 \cdot 1}{4 \cdot 3 \cdot 2 \cdot 1} = 30$

15.

$_4P_4 = \frac{4!}{(4-4)!} = \frac{4 \cdot 3 \cdot 2 \cdot 1}{0!} = 24$

17.

$_{22}C_4 = \frac{22!}{(22-4)! \, 4!} = 7315$

19.

$_7P_4 = \frac{7!}{(7-4)!} = \frac{7 \cdot 6 \cdot 5 \cdot 4 \cdot 3 \cdot 2 \cdot 1}{3 \cdot 2 \cdot 1} = 840$

21.

$_{10}P_6 = \frac{10!}{(10-6)!} = \frac{10 \cdot 9 \cdot 8 \cdot 7 \cdot 6 \cdot 5 \cdot 4 \cdot 3 \cdot 2 \cdot 1}{4 \cdot 3 \cdot 2 \cdot 1} = 151,200$

23.

$_{50}P_4 = \frac{50!}{(50-4)!} = \frac{50!}{46!} = 5,527,200$

25.

Same task: $_{12}C_4 = \frac{12!}{(12-4)! \, 4!} = 495$

Different tasks: $_{12}P_4 = \frac{12!}{(12-4)!} = 11,880$

27.

a. $\frac{5!}{3! \, 2!} = 10$ f. $\frac{3!}{3! \, 0!} = 1$

27. continued

b. $\frac{8!}{5!\,3!} = 56$ g. $\frac{3!}{0!\,3!} = 1$

c. $\frac{7!}{3!\,4!} = 35$ h. $\frac{9!}{2!\,7!} = 36$

d. $\frac{6!}{4!\,2!} = 15$ i. $\frac{12!}{10!\,2!} = 66$

e. $\frac{6!}{2!\,4!} = 15$ j. $\frac{4!}{1!\,3!} = 4$

29.

$_6P_3 = \frac{6!}{3!} = \frac{6\cdot5\cdot4\cdot3!}{3!} = 120$

31.

$_{10}C_4 = \frac{10!}{6!\,4!} = 210$

33.

$_{20}C_5 = \frac{20!}{15!\,5!} = 15{,}504$

35.

Different programs:

$_{18}C_{10} = \frac{18!}{(18-10)!\,10!} = 43{,}758$

Starting and ending with the same song:

$_{16}C_8 = \frac{16!}{(16-8)!\,8!} = 12{,}870$

37.

$_{12}C_4 = 495$

$_7C_2 \cdot {_5C_2} = 21 \cdot 10 = 210$

$_7C_2 \cdot {_5C_2} + {_7C_3} \cdot {_5C_1} + {_7C_4} =$

$21 \cdot 10 + 35 \cdot 5 + 35 =$

$210 + 175 + 35 = 420$

39.

The possibilities are CVV or VCV or VVV.

Assuming the same vowel can't be used twice in a "word":

$7 \cdot 5 \cdot 4 + 5 \cdot 7 \cdot 4 + 5 \cdot 4 \cdot 3 = 340$

Assuming the same vowel can be used twice in a "word":

$7 \cdot 5 \cdot 5 + 5 \cdot 7 \cdot 5 + 5 \cdot 5 \cdot 5 = 475$

41.

$_{10}C_2 \cdot {_{12}C_2} = \frac{10!}{8!\,2!} \cdot \frac{12!}{10!\,2!}$

$= 45 \cdot 66 = 2{,}970$

43.

There are $_7C_2 = 21$ tiles with unequal numbers and 7 tiles with equal numbers. Thus, the total number of tiles is 28.

45.

$_{11}C_7 = \frac{11!}{4!\,7!} = \frac{11\cdot10\cdot9\cdot8\cdot7!}{7!\cdot4\cdot3\cdot2\cdot1} = 330$

47.

$_{10}C_6 \cdot {_{12}C_6} = \frac{10!}{4!\,6!} \cdot \frac{12!}{6!\,6!} = 194{,}040$

49.

$_{20}C_8 = \frac{20!}{(20-8)!\,8!}$

$= \frac{20\cdot19\cdot18\cdot17\cdot16\cdot15\cdot14\cdot13\cdot12!}{12!\cdot8\cdot7\cdot6\cdot5\cdot4\cdot3\cdot2\cdot1} = 125{,}970$

51.

$_{20}P_5 = \frac{20!}{15!} = \frac{20\cdot19\cdot18\cdot17\cdot16\cdot15!}{15!} = 1{,}860{,}480$

53.

$_{17}C_2 = \frac{17!}{(17-2)!\,2!} = \frac{17\cdot16\cdot15!}{15!\cdot2\cdot1} = 136$

55.

$_5P_5 = \frac{5!}{0!} = \frac{5\cdot4\cdot3\cdot2\cdot1}{1} = 120$

57.

$_6C_3 \cdot {_5C_2} = \frac{6!}{3!\,3!} \cdot \frac{5!}{3!\,2!} = 200$

59.

$_8P_3 = \frac{8!}{5!} = \frac{8\cdot7\cdot6\cdot5\cdot4\cdot3\cdot2\cdot1}{5\cdot4\cdot3\cdot2\cdot1} = 336$

61.

$1 \cdot 2 \cdot 1 = 2$

$1 \cdot 3 \cdot 2 \cdot 1 = 6$

$1 \cdot (n-1) \cdot (n-2) \cdots 3 \cdot 2 \cdot 1 = (n-1)!$

63.

a. $_4C_1 \cdot 1 \cdot 1 \cdot 1 \cdot 1 = 4$

b. $_{10}C_1 \cdot {_4C_1} - {_4C_1} = 36$

c. $_{13}C_1 \cdot {_{12}C_1} \cdot {_4C_1} = 624$

d. $_{13}C_1 \cdot {_4C_3} \cdot {_{12}C_1} \cdot {_4C_2} = 3744$

EXERCISE SET 4-5

1.

$P(\text{2 face cards}) = \frac{12}{52} \cdot \frac{11}{51} = \frac{11}{221}$

3.

a. There are $_4C_3$ ways of selecting 3 women and $_7C_3$ total ways to select 3 people; hence, $P(\text{all women}) = \frac{_4C_3}{_7C_3} = \frac{4}{35}$.

b. There are $_3C_3$ ways of selecting 3 men; hence, $P(\text{all men}) = \frac{_3C_3}{_7C_3} = \frac{1}{35}$.

3. continued

c. There are $_3C_2$ ways of selecting 2 men and $_4C_1$ ways of selecting one woman; hence, P(2 men and 1 woman) $= \frac{_3C_2 \cdot _4C_1}{_7C_3}$ $= \frac{12}{35}$.

d. There are $_3C_1$ ways to select one man and $_4C_2$ ways of selecting two women; hence, P(1 man and 2 women) $= \frac{_3C_1 \cdot _4C_2}{_7C_3} = \frac{18}{35}$.

5.

a. P(all Republicans) =
$\frac{_{51}C_3 \cdot _{48}C_0 \cdot _1C_0}{_{100}C_3} = \frac{20,825}{161,700} = 0.129$

b. P(all Democrats) =
$\frac{_{51}C_0 \cdot _{48}C_3 \cdot _1C_0}{_{100}C_3} = \frac{17,296}{161,700} = 0.107$

c. P(one Republican, Democrat, and Independent) =
$\frac{_{51}C_1 \cdot _{48}C_1 \cdot _1C_1}{_{100}C_3} = \frac{2448}{161,700} = 0.015$

7.
$\frac{2}{50} \cdot \frac{1}{49} = \frac{1}{1225}$

9.

a. P(one of each) =
$\frac{_{18}C_1 \cdot _{10}C_1 \cdot _3C_1}{_{31}C_3} = \frac{540}{4495} = 0.120$

b. P(no Navy members) =
$\frac{_{21}C_3}{_{31}C_3} = \frac{1330}{4495} = 0.296$

c. P(three Marines) =
$\frac{_{18}C_3 \cdot _{10}C_0 \cdot _3C_0}{_{31}C_3} = \frac{816}{4495} = 0.182$

11.

a. P(red) $= \frac{_{11}C_2}{_{19}C_2} = \frac{55}{171} = 0.3216$

b. P(black) $= \frac{_8C_2}{_{19}C_2} = \frac{28}{171} = 0.1637$

c. P(unmatched) $= \frac{_{11}C_1 \cdot _8C_1}{_{19}C_2} = \frac{88}{171} = 0.5146$

d. It probably got lost in the wash!

13.
There are $6^3 = 216$ ways of tossing three dice, and there are 15 ways of getting a sum of 7; i.e., (1, 1, 5), (1, 5, 1), (5, 1, 1), (1, 2, 4), etc. Hence the probability of rolling a sum of 7 is $\frac{15}{216} = \frac{5}{72}$.

15.
There are $5! = 120$ ways to arrange 5 washers in a row and 2 ways to have them in correct order, small to large or large to small; hence, the probability is $\frac{2}{120} = \frac{1}{60}$.

17.
P(berries are produced) = P(either 1 or 2 males)
P(1 or 2 males) $= \frac{_8C_2 \cdot _4C_1}{_{12}C_3} + \frac{_8C_1 \cdot _4C_2}{_{12}C_3}$
$= 0.509 + 0.218 = 0.727$

REVIEW EXERCISES - CHAPTER 4

1.
a. $\frac{1}{6}$ b. $\frac{2}{3}$ c. $\frac{3}{6} = \frac{1}{2}$

3.
a. P(not used for taxes) = P(virus or other)
P(virus or other) $= \frac{5}{10} + \frac{2}{10} = 0.7$

b. P(taxes or other use) $= \frac{3}{10} + \frac{2}{10} = 0.5$

5.
P(preferred juice) $= \frac{13}{60}$

7.
P(either air-conditioning or CD player)
$= 0.5 + 0.37 - 0.06 = 0.81$
P(neither air-conditioning nor CD)
$= 1 - 0.81 = 0.19$

9.
$0.80 + 0.30 - 0.12 = 0.98$

11.
P(enrolled in an online course) $= \frac{1}{6}$ or 0.167
a. P(all 5 took an online course) $= (\frac{1}{6})^5 = 0.0001$

b. P(none took an online course) $= (\frac{5}{6})^5 = 0.402$

c. P(at least one took an online course)
$= 1 - $ P(none took an online course)
$= 1 - (\frac{5}{6})^5 = 0.598$

13.
a. $\frac{26}{52} \cdot \frac{25}{51} \cdot \frac{24}{50} = \frac{2}{17}$

b. $\frac{13}{52} \cdot \frac{12}{51} \cdot \frac{11}{50} = \frac{33}{2550} = \frac{11}{850}$

c. $\frac{4}{52} \cdot \frac{3}{51} \cdot \frac{2}{50} = \frac{1}{5525}$

15.
Total number of movie releases = 1384
a. P(European) = $\frac{834}{1384}$ = 0.603

b. P(US) = $\frac{471}{1384}$ = 0.340

c. P(German or French) = $\frac{316}{1384} + \frac{132}{1384}$
= $\frac{448}{1384}$ or 0.324

d. P(German | European)
= $\frac{\text{P(European and German)}}{\text{P(European)}} = \frac{\frac{316}{1384}}{\frac{834}{1384}}$ = 0.379

17.

P(disease) = 0.025 + 0.375 = 0.4

19.
P(NC | C) = $\frac{\text{P(NC and C)}}{\text{P(C)}} = \frac{0.37}{0.73}$ = 0.51

21.
$\frac{0.43}{0.75}$ = 0.573 or 57.3%

23.

	<4 yrs HS	HS	College	Total
Smoker	6	14	19	39
Non-Smoker	18	7	25	50
Total	24	21	44	89

a. There are 44 college graduates and 19 of them smoke; hence, the probability is $\frac{19}{44}$.

b. There are 24 people who did not graduate from high school, 6 of whom do not smoke; hence, the probability is
$\frac{6}{24} = \frac{1}{4}$.

25.
P(at least one household has no DVD player)
= 1 − P(none have no DVD player)
= 1 − P(all 6 have DVD players)
= 1 − (0.81)^6 = 0.718

27.
If repetitions are allowed:
26 · 26 · 26 · 10 · 10 · 10 = 175, 760, 000

If repetitions are not allowed:
$_{26}P_3 \cdot {_{10}}P_4 = \frac{26 \cdot 25 \cdot 24 \cdot 23!}{23!} \cdot \frac{10 \cdot 9 \cdot 8 \cdot 7 \cdot 6!}{6!}$
= 78,624,000

If repetitions are allowed in the letters but not in the digits:
26 · 26 · 26 · $_{10}P_4$ = 88,583,040

29.
$_5C_3 \cdot {_7}C_4 = \frac{5!}{2! \, 3!} \cdot \frac{7!}{3! \, 4!} = 10 \cdot 35 = 350$

31.
$_{10}C_2 = \frac{10!}{8! \, 2!} = 45$

33.
100!

35.
$_{12}C_4 = \frac{12!}{8! \, 4!} = \frac{12 \cdot 11 \cdot 10 \cdot 9 \cdot 8!}{4 \cdot 3 \cdot 2 \cdot 1 \cdot 8!} = 495$

37.
$_{20}C_5 = \frac{20!}{15! \, 5!} = \frac{20 \cdot 19 \cdot 18 \cdot 17 \cdot 16 \cdot 15!}{15! \, 5 \cdot 4 \cdot 3 \cdot 2 \cdot 1} = 15,504$

39.
Total number of outcomes:
26 · 26 · 26 · 10 · 10 · 10 · 10 = 175,760,000

Total number of ways for USA followed by a number divisible by 5:
1 · 1 · 1 · 10 · 10 · 10 · 2 = 2000

Hence P = $\frac{2000}{175,760,000}$ = 0.0000114

41.
Total number of territories = 45

P(3 French or 3 UK or 3 US) = $\frac{_{16}C_3}{_{45}C_3} + \frac{_{15}C_3}{_{45}C_3} + \frac{_{14}C_3}{_{45}C_3}$

= $\frac{560}{14,190} + \frac{455}{14,190} + \frac{364}{14,190}$

= $\frac{1379}{14,190}$ = 0.097

43.

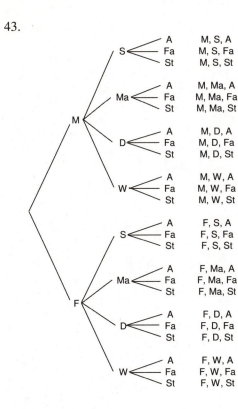

	S	A	M, S, A
		Fa	M, S, Fa
		St	M, S, St
	Ma	A	M, Ma, A
		Fa	M, Ma, Fa
		St	M, Ma, St
M	D	A	M, D, A
		Fa	M, D, Fa
		St	M, D, St
	W	A	M, W, A
		Fa	M, W, Fa
		St	M, W, St
	S	A	F, S, A
		Fa	F, S, Fa
		St	F, S, St
	Ma	A	F, Ma, A
		Fa	F, Ma, Fa
		St	F, Ma, St
F	D	A	F, D, A
		Fa	F, D, Fa
		St	F, D, St
	W	A	F, W, A
		Fa	F, W, Fa
		St	F, W, St

CHAPTER 4 QUIZ

1. False, subjective probability can be used when other types of probabilities cannot be found.
2. False, empirical probability uses frequency distributions.
3. True
4. False, $P(A \text{ or } B) = P(A) + P(B) - P(A \text{ and } B)$
5. False, the probabilities can be different.
6. False, complementary events cannot occur at the same time.
7. True
8. False, order does not matter in combinations.
9. b
10. b and d
11. d
12. b
13. c
14. b
15. d
16. b
17. b
18. sample space
19. zero and one
20. zero
21. one
22. mutually exclusive

23. a. $\frac{4}{52} = \frac{1}{13}$ c. $\frac{16}{52} = \frac{4}{13}$

b. $\frac{4}{52} = \frac{1}{13}$

24. a. $\frac{13}{52} = \frac{1}{4}$ d. $\frac{4}{52} = \frac{1}{13}$

b. $\frac{4+13-1}{52} = \frac{4}{13}$ e. $\frac{26}{52} = \frac{1}{2}$

c. $\frac{1}{52}$

25. a. $\frac{12}{31}$ c. $\frac{27}{31}$

b. $\frac{12}{31}$ d. $\frac{24}{31}$

26. a. $\frac{11}{36}$ d. $\frac{1}{3}$

b. $\frac{5}{18}$ e. 0

c. $\frac{11}{36}$ f. $\frac{11}{12}$

27. $(0.75 - 0.16) + (0.25 - 0.16) = 0.68$

28. $(0.3)^5 = 0.002$

29. a. $\frac{26}{52} \cdot \frac{25}{51} \cdot \frac{24}{50} \cdot \frac{23}{49} \cdot \frac{22}{48} = \frac{253}{9996}$

b. $\frac{13}{52} \cdot \frac{12}{51} \cdot \frac{11}{50} \cdot \frac{10}{49} \cdot \frac{9}{48} = \frac{33}{66,640}$

c. 0

30. $\frac{0.35}{0.65} = 0.54$

31. $\frac{0.16}{0.3} = 0.53$

32. $\frac{0.57}{0.7} = 0.81$

33. $\frac{0.028}{0.5} = 0.056$

34. a. $\frac{1}{2}$ b. $\frac{3}{7}$

35. $1 - (0.45)^6 = 0.99$

36. $1 - (\frac{5}{6})^4 = 0.518$

37. $1 - (0.15)^6 = 0.9999886$

38. 2,646

39. 40,320

40. 1,365

41. 1,188,137,600; 710,424,000

42. 720

43. 33,554,432

44. 56

45. $\frac{1}{4}$

46. $\frac{3}{14}$

47. $\frac{12}{55}$

48.

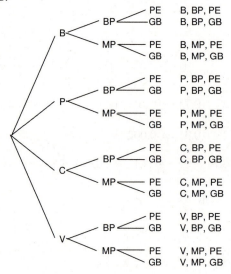

Note: Answers may vary due to rounding, TI-83's or computer programs.

EXERCISE SET 5-1

1.
A random variable is a variable whose values are determined by chance. Examples will vary.

3.
The number of commercials a radio station plays during each hour.
The number of times a student uses his or her calculator during a mathematics exam.
The number of leaves on a specific type of tree.

5.
A probability distribution is a distribution which consists of the values a random variable can assume along with the corresponding probabilities of these values.

7.
No; probabilities cannot be negative and the sum of the probabilities is not 1.

9.
Yes

11.
No, probability values cannot be greater than 1.

13.
Discrete

15.
Continuous

17.
Discrete

19.

X	0	1	2	3
P(X)	$\frac{6}{15}$	$\frac{5}{15}$	$\frac{3}{15}$	$\frac{1}{15}$

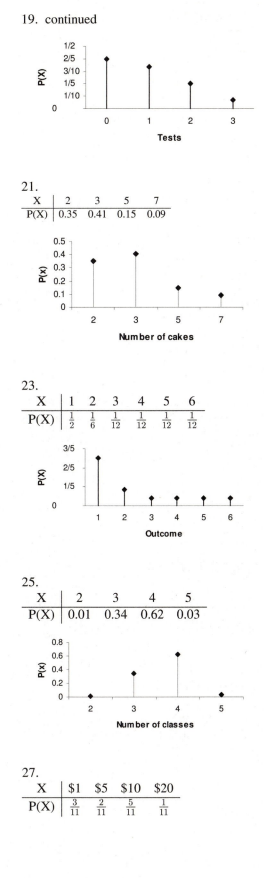

19. continued

21.

X	2	3	5	7
P(X)	0.35	0.41	0.15	0.09

23.

X	1	2	3	4	5	6
P(X)	$\frac{1}{2}$	$\frac{1}{6}$	$\frac{1}{12}$	$\frac{1}{12}$	$\frac{1}{12}$	$\frac{1}{12}$

25.

X	2	3	4	5
P(X)	0.01	0.34	0.62	0.03

27.

X	$1	$5	$10	$20
P(X)	$\frac{3}{11}$	$\frac{2}{11}$	$\frac{5}{11}$	$\frac{1}{11}$

27. continued

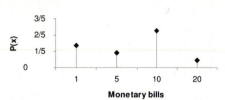

29.

X	1	2	3	4
P(X)	$\frac{1}{4}$	$\frac{1}{4}$	$\frac{3}{8}$	$\frac{1}{8}$

31.

X	1	2	3
P(X)	$\frac{1}{6}$	$\frac{1}{3}$	$\frac{1}{2}$

Yes.

33.

X	3	4	7
P(X)	$\frac{3}{6}$	$\frac{4}{6}$	$\frac{7}{6}$

No, the sum of the probabilities is greater than one and $P(7) = \frac{7}{6}$ which is also greater than 1.

35.

X	1	2	4
P(X)	$\frac{1}{7}$	$\frac{2}{7}$	$\frac{4}{7}$

Yes.

EXERCISE SET 5-2

1.

X	0	1	2	3
P(X)	0.90	0.05	0.03	0.02

$\mu = \sum X \cdot P(X) = 0(0.9) + 1(0.05) + 2(0.03) + 3(0.02) = 0.17$

$\sigma^2 = \sum X^2 \cdot P(X) - \mu^2 = [0^2(0.9) + 1^2(0.05) + 2^2(0.03) + 3^2(0.02)] - 0.17^2 = 0.321$

$\sigma = \sqrt{0.321} = 0.567$

1. continued

X	P(X)	$X \cdot P(X)$	$X^2 \cdot P(X)$
0	0.90	0	0
1	0.05	0.05	0.05
2	0.03	0.06	0.12
3	0.02	0.06	0.18
		$\mu = 0.17$	0.35

3.
$\mu = \sum X \cdot P(X) = 0(0.18) + 1(0.44) + 2(0.27) + 3(0.08) + 4(0.03) = 1.34$ or 1.3

$\sigma^2 = \sum X^2 \cdot P(X) - \mu^2 = [0^2(0.18) + 1^2(0.44) + 2^2(0.27) + 3^2(0.08) + 4^2(0.03)] - 1.34^2 = 0.92$ or 0.9

$\sigma = \sqrt{0.92} = 0.96$ or 1

No, on average each person has about one credit card.

X	P(X)	$X \cdot P(X)$	$X^2 \cdot P(X)$
0	0.18	0	0
1	0.44	0.44	0.44
2	0.27	0.54	1.08
3	0.08	0.24	0.72
4	0.03	0.12	0.48
		$\mu = 1.34$	2.72

5.
$\mu = \sum X \cdot P(X) = 4(0.4) + 5(0.3) + 6(0.1) + 8(0.15) + 10(0.05) = 5.4$

$\sigma^2 = \sum X^2 \cdot P(X) - \mu^2 = [4^2(0.4) + 5^2(0.3) + 6^2(0.1) + 8^2(0.15) + 10^2(0.05)] - 5.4^2 = 2.94$

$\sigma = \sqrt{2.94} = 1.71$

X	P(X)	$X \cdot P(X)$	$X^2 \cdot P(X)$
4	0.40	1.6	6.4
5	0.30	1.5	7.5
6	0.10	0.6	3.6
8	0.15	1.2	9.6
10	0.05	0.5	5.0
		$\mu = 5.4$	32.1

P(selling 6 or more contracts) = 0.3

7.
$\mu = \sum X \cdot P(X) = 5(0.2) + 6(0.25) + 7(0.38) + 8(0.10) + 9(0.07) = 6.59$ or 6.6

7. continued
$\sigma^2 = \sum X^2 \cdot P(X) - \mu^2 = [5^2(0.2) + 6^2(0.25) + 7^2(0.38) + 8^2(0.10) + 9^2(0.07) - 6.59^2 = 1.2619$ or 1.3

$\sigma = \sqrt{1.2619} = `1.123$ or 1.1

X	P(X)	X · P(X)	X² · P(X)
5	0.20	1.00	5.00
6	0.25	1.50	9.00
7	0.38	2.66	18.62
8	0.10	0.80	6.40
9	0.07	0.63	5.67
		$\mu = 6.59$	44.69

9.
$\mu = \sum X \cdot P(X) = 6(0.15) + 8(0.3) + 10(0.35) + 12(0.1) + 14(0.1) = 9.4$

$\sigma^2 = \sum X^2 \cdot P(X) - \mu^2 = [6^2(0.15) + 8^2(0.3) + 10^2(0.35) + 12^2(0.1) + 14^2(0.1)] - 9.4^2 = 5.24$

$\sigma = \sqrt{5.24} = 2.289$

X	P(X)	X · P(X)	X² · P(X)
6	0.15	0.9	5.4
8	0.30	2.4	19.2
10	0.35	3.5	35.0
12	0.10	1.2	14.4
14	0.10	1.4	19.6
		$\mu = 9.4$	93.6

$P(< 8 \text{ or } > 12) = 0.15 + 0.10 = 0.25$

11.
$E(X) = \sum X \cdot P(X) = \$300(0.998) - \$19,700(0.002) = \$260$

13.
$E(X) = \sum X \cdot P(X) = \$5.00(\frac{1}{6}) = \$0.83$
He should pay about $0.83.

15.
$E(X) = \sum X \cdot P(X) = \$1000(\frac{1}{1000}) + \$500(\frac{1}{1000}) + \$100(\frac{5}{1000}) - \$3.00 = -\$1.00$

Alternate Solution:
$E(X) = 997(\frac{1}{1000}) + 497(\frac{1}{1000}) + 97(\frac{5}{1000}) - 3(\frac{993}{1000}) = -\$1.00$

17.
$E(X) = \sum X \cdot P(X) = \$500(\frac{1}{1000}) - \$1.00 = -\$0.50$

If 123 is boxed:
$E(X) = \$499(\frac{1}{1000}) - 1(\frac{999}{1000}) = -\$0.50$

There are 6 possibilities when a number with all different digits is boxed, $(3 \cdot 2 \cdot 1 = 6)$. Hence,
$\$80.00 \cdot \frac{6}{1000} - \$1.00 = \$0.48 - \$1.00 = -\$0.52$

Alternate Solution:
$E(X) = 79(\frac{6}{1000}) - 1(\frac{994}{1000}) = -\$0.52$

19.
a. $P(\text{red}) = \frac{18}{38}(\$1) + \frac{20}{38}(-\$1) = -\$0.05$

b. $P(\text{even}) = \frac{18}{38}(\$1) + \frac{20}{38}(-\$1) = -\$0.05$

c. $P(00) = \frac{1}{38}(\$1) + \frac{37}{38}(-\$1) = -\$0.05$

d. $P(\text{any single number}) = \frac{1}{38}(\$1) + \frac{37}{38}(-\$1) = -\$0.05$

e. $P(0 \text{ or } 00) = \frac{2}{38}(\$17) + \frac{36}{38}(-\$1) = -\$0.05$

21.
The expected value for a single die is 3.5, and since 3 die are rolled, the expected value is $3(3.5) = 10.5$

23.
Answers will vary.

25.
Answers will vary.

EXERCISE SET 5-3

1.
a. Yes
b. Yes
c. Yes
d. No, there are more than two outcomes.
e. No, there are more than two outcomes.
f. Yes
g. Yes
h. Yes
i. No, there are more than two outcomes.
j. Yes

2.
a. 0.420
b. 0.346
c. 0.590
d. 0.251
e. 0.000
f. 0.250
g. 0.418
h. 0.176
i. 0.246

3.
a. $P(X) = \frac{n!}{(n-X)!\,X!} \cdot p^X \cdot q^{n-X}$

$P(X) = \frac{6!}{3!\cdot 3!} \cdot (0.03)^3 (0.97)^3 = 0.0005$

b. $P(X) = \frac{4!}{2!\cdot 2!} \cdot (0.18)^2 \cdot (0.82)^2 = 0.131$

c. $P(X) = \frac{5!}{2!\cdot 3!} = (0.63)^3 \cdot (0.37)^2 = 0.342$

d. $P(X) = \frac{9!}{9!\cdot 0!} \cdot (0.42)^0 \cdot (0.58)^9 = 0.007$

e. $P(X) = \frac{10!}{5!\cdot 5!} \cdot (0.37)^5 \cdot (0.63)^5 = 0.173$

5.
$n = 20, p = 0.5, X \geq 15$
$P(X) = 0.015 + 0.005 + 0.001 = 0.021$
No, it's only about a 2% chance.

7.
a. $n = 8, p = 0.77, X = 8$

$P(X) = \frac{8!}{0!\,8!}(0.77)^8 (0.23)^0 = 0.124$

b. $n = 8, P = 0.77, X = 5, 6, 7,$ or 8

$P(X) = \frac{8!}{3!5!}(0.77)^5 (0.23)^3 +$

$\frac{8!}{2!6!}(0.77)^6 (0.23)^2 + \frac{8!}{1!7!}(0.77)^7 (0.23)^1 +$

$\frac{8!}{0!8!}(0.77)^8 (0.23)^0 = 0.184 + 0.309 + 0.295$

$+ 0.124 = 0.912$

c. $n = 8, p = 0.77, X = 3$

$P(X) = \frac{8!}{5!3!}(0.77)^3 (0.23)^5 = 0.017$

9.
$n = 7, p = 0.75, X = 0, 1, 2, 3$

9. continued
$P(X) = \frac{7!}{7!\,0!}(0.75)^0 (0.25)^7 +$

$\frac{7!}{6!\,1!}(0.75)^1 (0.25)^6 + \frac{7!}{5!\,2!}(0.75)^2 (0.25)^5 +$

$\frac{7!}{4!\,3!}(0.75)^3 (0.25)^4 = 0.071$

11.
$n = 5, p = 0.40$
a. $X = 2, P(X) = 0.346$

b. $X = 0, 1, 2,$ or 3 people
$P(X) = 0.078 + 0.259 + 0.346 + 0.230$
$= 0.913$

c. $X = 2, 3, 4,$ or 5 people
$P(X) = 0.346 + 0.230 + 0.077 + 0.01$
$= 0.663$

d. $X = 0, 1,$ or 2 people
$P(X) = 0.683$

13.
a. $n = 10, p = 0.53, X = 5$

$P(X) = \frac{10!}{5!5!}(0.53)^5 (0.47)^5 = 0.2417$

b. $n = 10, p = 0.47, X \geq 5$

$P(X) = \frac{10!}{5!5!}(0.47)^5 (0.53)^5 +$

$\frac{10!}{6!4!}(0.47)^6 (0.53)^4 + \frac{10!}{7!3!}(0.47)^7 (0.53)^3 +$

$\frac{10!}{8!2!}(0.47)^8 (0.53)^2 + \frac{10!}{9!1!}(0.47)^9 (0.53)^1 +$

$\frac{10!}{10!0!}(0.47)^{10} (0.53)^0$

$P(X \geq 5) = 0.548$

c. $n = 10, p = 0.53, X < 5$

$P(X) = \frac{10!}{5!5!}(0.53)^5 (0.47)^5 +$

$\frac{10!}{4!6!}(0.53)^4 (0.47)^6 + \frac{10!}{3!7!}(0.53)^3 (0.47)^7 +$

$\frac{10!}{2!8!}(0.53)^2 (0.47)^8 + \frac{10!}{1!9!}(0.53)^1 (0.47)^9$

$\frac{10!}{0!10!}(0.53^0)(0.47)^{10} = 0.306$

14.
a. $\mu = 100(0.75) = 75$
$\sigma^2 = 100(0.75)(0.25) = 18.75$ or 18.8
$\sigma = \sqrt{18.75} = 4.33$ or 4.3

b. $\mu = 300(0.3) = 90$
$\sigma^2 = 300(0.3)(0.7) = 63$
$\sigma = \sqrt{63} = 7.94$ or 7.9

c. $\mu = 20(0.5) = 10$
$\sigma^2 = 20(0.5)(0.5) = 5$
$\sigma = \sqrt{5} = 2.236$ or 2.2

d. $\mu = 10(0.8) = 8$
$\sigma^2 = 10(0.8)(0.2) = 1.6$
$\sigma = \sqrt{1.6} = 1.265$ or 1.3

e. $\mu = 1000(0.1) = 100$
$\sigma^2 = 1000(0.1)(0.9) = 90$
$\sigma = \sqrt{90} = 9.49$ or 9.5

f. $\mu = 500(0.25) = 125$
$\sigma^2 = 500(0.25)(0.75) = 93.75$ or 93.8
$\sigma = \sqrt{93.75} = 9.68$ or 9.7

g. $\mu = 50(\frac{2}{5}) = 20$
$\sigma^2 = 50(\frac{2}{5})(\frac{3}{5}) = 12$
$\sigma = \sqrt{12} = 3.464$ or 3.5

h. $\mu = 36(\frac{1}{6}) = 6$
$\sigma^2 = 36(\frac{1}{6})(\frac{5}{6}) = 5$
$\sigma = \sqrt{5} = 2.236$ or 2.2

15.
$n = 800$, $p = 0.01$
$\mu = 800(0.01) = 8$
$\sigma^2 = 800(0.01)(0.99) = 7.9$
$\sigma = \sqrt{7.92} = 2.8$

17.
$n = 300$, $p = 0.03$
$\mu = 300(0.03) = 9$
$\sigma^2 = 300(0.03)(0.97) = 8.73$
$\sigma = \sqrt{8.73} = 2.95$

19.
$n = 1000$, $p = 0.21$
$\mu = 1000(0.21) = 210$
$\sigma^2 = 1000(0.21)(0.79) = 165.9$
$\sigma = \sqrt{165.9} = 12.9$

21.
$n = 18$, $p = 0.25$, $X = 5$
$P(X) = \frac{18!}{13!\,5!}(0.25)^5(0.75)^{13} = 0.199$

23.
$n = 10$, $p = \frac{1}{3}$, $X = 0, 1, 2, 3$

23. continued
$P(X) = \frac{10!}{10!\,0!}(\frac{1}{3})^0(\frac{2}{3})^{10} + \frac{10!}{9!\,1!}(\frac{1}{3})^1(\frac{2}{3})^9$
$+ \frac{10!}{8!\,2!}(\frac{1}{3})^2(\frac{2}{3})^8 + \frac{10!}{7!\,3!}(\frac{1}{3})^3(\frac{2}{3})^7 = 0.559$

25.
$n = 20$, $p = 0.58$, $X = 12$
$P(X) = \frac{20!}{8!\,12!}(0.58)^{12}(0.42)^8 = 0.177$

27.
$n = 7$, $p = 0.14$, $X = 2$ or 3
$P(X) = \frac{7!}{5!\,2!}(0.14)^2(0.86)^5 +$
$\frac{7!}{4!\,3!}(0.14)^3(0.86)^4 = 0.246$

29.

X	0	1	2	3
P(X)	0.125	0.375	0.375	0.125

EXERCISE SET 5-4

1.
a. $P(M) = \frac{6!}{3!\,2!\,1!}(0.5)^3(0.3)^2(0.2)^1 = 0.135$

b. $P(M) = \frac{5!}{1!\,2!\,2!}(0.3)^1(0.6)^2(0.1)^2 = 0.0324$

c. $P(M) = \frac{4!}{1!\,1!\,2!}(0.8)^1(0.1)^1(0.1)^2 = 0.0096$

d. $P(M) = \frac{3!}{1!\,1!\,1!}(0.5)^1(0.3)^1(0.2)^1 = 0.18$

e. $P(M) = \frac{5!}{1!\,3!\,1!}(0.7)^1(0.2)^3(0.1)^1 = 0.0112$

3.
$P(M) =$
$\frac{12!}{2!\,2!\,2!\,2!\,2!\,2!}(0.13)^2(0.13)^2(0.14)^2(0.16)^2(0.2)^2$

$(0.24)^2 = 0.00247$

5.
$P(M) = \frac{4!}{2!\,1!\,1!}(\frac{1}{6})^2(\frac{1}{6})^1(\frac{1}{6})^1 = \frac{1}{108}$

7.
a. $P(5; 4) = 0.1563$
b. $P(2; 4) = 0.1465$
c. $P(6; 3) = 0.0504$
d. $P(10; 7) = 0.071$
e. $P(9; 8) = 0.1241$

9.
$p = \frac{1}{20,000} = 0.00005$
$\lambda = n \cdot p = 80,000(0.00005) = 4$
a. $P(0; 4) = 0.0183$
b. $P(1; 4) = 0.0733$
c. $P(2; 4) = 0.1465$

9. continued

d. P(3 or more; 4) = 1 − [P(0; 4) + P(1; 4)
+ P(2; 4)]
= 1 − (0.0183 + 0.0733 + 0.1465)
= 0.7619

11.
$p = \frac{5}{1000} = \frac{1}{200}$
$\lambda = n \cdot p = (250) \cdot (\frac{1}{200}) = 1.25$
P(at least 2 orders) =
1 − [P(0 orders + P(1 order)]
$= 1 - [\frac{e^{-1.25}(1.25)^0}{0!} + \frac{e^{-1.25}(1.25)^1}{1!}]$

= 1 − (0.2865 + 0.3581) = 0.3554

13.
$\lambda = 200(0.015) = 3$
P(0; 3) = 0.0498

15.
P(5; 4) = 0.1563

17.
P(one from each
class) $= \frac{_5C_1 \cdot _4C_1 \cdot _5C_1 \cdot _7C_1}{_{21}C_4} = \frac{700}{5985} = 0.117$

19.
P(2 jazz and 1
classical) $= \frac{_{10}C_2 \cdot _4C_1 \cdot _2C_0}{_{16}C_3} = \frac{180}{560} = 0.321$

21.
P(at least 1 defective) = 1 − P(0 defectives)
a = 6, b = 18, n = 3, X = 0

$P(0) = \frac{_6C_0 \cdot _{18}C_3}{_{24}C_3} = \frac{102}{253} = 0.403$
P(at least 1 defective) = 1 − 0.403 = 0.597

REVIEW EXERCISES - CHAPTER 5

1.
Yes

3.
No, the sum of the probabilities is greater
than 1.

5.

X	0	1	2	3	4	5
P(X)	0.27	0.28	0.2	0.15	0.08	0.02

5. continued

Number of applications

a. P(2 or 3
applications) = 0.2 + 0.15 = 0.35

b.
$\mu = 0(0.27) + 1(0.28) + 2(0.2) + 3(0.15)$
$+ 4(0.08) + 5(0.02) = 1.55$

$\sigma^2 = [0^2(0.27) + 1^2(0.28) + 2^2(0.2) +$
$3^2(0.15) + 4^2(0.08) + 5^2(0.02)] - 1.55^2$
$= 1.8075$

$\sigma = \sqrt{1.8075} = 1.3444$

X	P(X)	X · P(X)	X² · P(X)
0	0.27	0	0
1	0.28	0.28	0.28
2	0.2	0.40	0.80
3	0.15	0.45	1.35
4	0.08	0.32	1.28
5	0.02	0.10	0.50
		μ = 1.55	4.21

7.

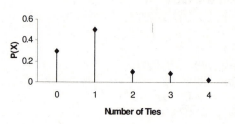
Number of Ties

9.
$\mu = \sum X \cdot P(X) = 5(0.14) + 6(0.21) +$
$7(0.24) + 8(0.18) + 9(0.16) + 10(0.07)$
$= 7.22$

$\sigma^2 = \sum X^2 \cdot P(X) - \mu^2 = [5^2(0.14) +$
$6^2(0.21) + 7^2(0.24) + 8^2(0.18) +$
$9^2(0.16) + 10^2(0.07)] - 7.22^2$
$= 2.1716$

$\sigma = \sqrt{2.1716} = 1.47$

9. continued

X	P(X)	X · P(X)	$X^2 \cdot P(X)$
5	0.14	0.7	3.5
6	0.21	1.26	7.56
7	0.24	1.68	11.76
8	0.18	1.44	11.52
9	0.16	1.44	12.96
10	0.07	0.7	7.00
		$\mu = 7.22$	54.3

11.

$\mu = \sum X \cdot P(X) = 22(0.08) + 23(0.19) +$
$24(0.36) + 25(0.25) + 26(0.07) + 27(0.05)$
$= 24.19$ or 24.2

$\sigma^2 = \sum X^2 \cdot P(X) - \mu^2 = [22^2(0.08) +$
$23^2(0.19) + 24^2(0.36) + 25^2(0.25) +$
$26^2(0.07) + 27^2(0.05)] - 24.19^2 = 1.4539$
or 1.5

$\sigma = \sqrt{1.4539} = 1.206$ or 1.2

X	P(X)	X · P(X)	$X^2 \cdot P(X)$
22	0.08	1.76	38.72
23	0.19	4.37	100.51
24	0.36	8.64	207.36
25	0.25	6.25	156.25
26	0.07	1.82	47.32
27	0.05	1.35	36.45
	$\mu =$	24.19	586.61

13.

Let x = cost to play the game
$P(ace) = \frac{1}{13}$ $P(face\ card) = \frac{3}{13}$
$P(2 - 10) = \frac{9}{13}$
For a fair game, E(X) = 0.
$0 = -20(\frac{1}{13}) + 10(\frac{3}{13}) + 2(\frac{9}{13}) - x$
$x = \$2.15$

15.

n = 12, p = 0.3
a. P(X = 8) = 0.008
b. P(X < 5) = 0.724
c. P(X ≥ 10) = 0.0002
d. P(4 < x ≤ 9) = 0.2761

17.

$\mu = n \cdot p = 150(0.8) = 120$
$\sigma^2 = n \cdot p \cdot q = 150(0.8)(0.2) = 24$
$\sigma = \sqrt{24} = 4.90$

19.

n = 8, p = 0.25

19. continued

$P(X \leq 3) = \frac{8!}{8!\ 0!}(0.25)^0(0.75)^8 +$
$\frac{8!}{7!\ 1!}(0.25)^1(0.75)^7 + \frac{8!}{6!\ 2!}(0.25)^2(0.75)^6 +$
$\frac{8!}{5!\ 3!}(0.25)^3(0.75)^5 = 0.8862$ or 0.886

21.

n = 20, p = 0.75, X = 16
P(16 have eaten pizza for breakfast) =

$\frac{20!}{4!\ 16!}(0.75)^{16}(0.25)^4 = 0.1897$ or 0.190

23.

$P(M) = \frac{10!}{5!\ 3!\ 1!\ 1!}(0.46)^5(0.41)^3(0.09)^1(0.03)^1$
$= 0.0193$

25.

$P(M) = \frac{10!}{5!\ 3!\ 2!}(0.50)^5(0.40)^3(0.10)^2 = 0.050$

27.

a. P(6 or more; 6) = 1 − P(5 or less; 6)
 = 1 − (0.0025 + 0.0149 + 0.0446 +
0.0892 + 0.1339 + 0.1606) = 0.5543
b. P(4 or more; 6) = 1 − P(3 or less; 6)
 = 1 − (0.0025 + 0.0149 + 0.0446 +
0.0892) = 0.8488
c. P(5 or less; 6) = P(0; 6) + ... + P(6; 6)
 = 0.4457

29.

a = 13, b = 39, n = 5, X = 2
$P(2) = \frac{_{13}C_2 \cdot _{39}C_3}{_{52}C_5} = \frac{9,139}{33,320} = 0.27$

31.

P(1 vegetable & 2 fruit) =

$\frac{_{10}C_0 \cdot _8C_1 \cdot _8C_2}{_{26}C_3} = \frac{224}{2600} = 0.0862$

CHAPTER 5 QUIZ

1. True
2. False, it is a discrete random variable.
3. False, the outcomes must be independent.
4. True
5. Chance
6. $\mu = n \cdot p$
7. One
8. c
9. c
10. d
11. No, the sum of the probabilities is greater than one.
12. Yes
13. Yes

Chapter 5 - Discrete Probability Distributions

14. Yes

15.

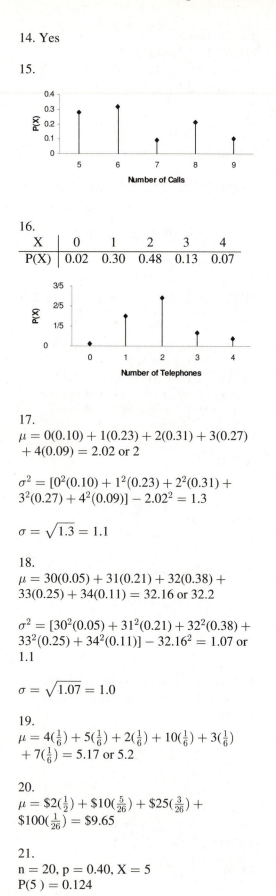

16.

X	0	1	2	3	4
P(X)	0.02	0.30	0.48	0.13	0.07

17.
$\mu = 0(0.10) + 1(0.23) + 2(0.31) + 3(0.27) + 4(0.09) = 2.02$ or 2

$\sigma^2 = [0^2(0.10) + 1^2(0.23) + 2^2(0.31) + 3^2(0.27) + 4^2(0.09)] - 2.02^2 = 1.3$

$\sigma = \sqrt{1.3} = 1.1$

18.
$\mu = 30(0.05) + 31(0.21) + 32(0.38) + 33(0.25) + 34(0.11) = 32.16$ or 32.2

$\sigma^2 = [30^2(0.05) + 31^2(0.21) + 32^2(0.38) + 33^2(0.25) + 34^2(0.11)] - 32.16^2 = 1.07$ or 1.1

$\sigma = \sqrt{1.07} = 1.0$

19.
$\mu = 4(\frac{1}{6}) + 5(\frac{1}{6}) + 2(\frac{1}{6}) + 10(\frac{1}{6}) + 3(\frac{1}{6}) + 7(\frac{1}{6}) = 5.17$ or 5.2

20.
$\mu = \$2(\frac{1}{2}) + \$10(\frac{5}{26}) + \$25(\frac{3}{26}) + \$100(\frac{1}{26}) = \$9.65$

21.
n = 20, p = 0.40, X = 5
P(5) = 0.124

22.
n = 20, p = 0.60
a. P(15) = 0.075

b. P(10, 11, ..., 20) = 0.872

c. P(0, 1, 2, 3, 4, 5) = 0.125

23.
n = 300, p = 0.80
$\mu = 300(0.80) = 240$
$\sigma^2 = 300(0.80)(0.20) = 48$
$\sigma = \sqrt{48} = 6.9$

24.
n = 75, p = 0.12
$\mu = 75(0.12) = 9$
$\sigma^2 = 75(0.12)(0.88) = 7.9$
$\sigma = \sqrt{7.9} = 2.8$

25.
$P(M) = \frac{30!}{15!\,8!\,5!\,2!}(0.5)^{15}(0.3)^8(0.15)^5(0.05)^2$

P(M) = 0.0080

26.
$P(M) = \frac{16!}{9!\,4!\,3!}(0.88)^9(0.08)^4(0.04)^3$

P(M) = 0.0003

27.
$P(M) = \frac{12!}{5!\,4!\,3!}(0.45)^5(0.35)^4(0.2)^3$

P(M) = 0.061

28.
$\lambda = 100(0.08) = 8$, X = 6
P(6; 8) = 0.122

29.
$\lambda = 8$
a. $P(X \geq 8; 8) = 0.1396 + \ldots + 0.0001 = 0.5471$

b. $P(X \geq 3; 8) = 1 - P(0, 1,$ or 2 calls$)$
$= 1 - (0.0003 + 0.0027 + 0.0107)$
$= 1 - 0.0137 = 0.9863$

c. $P(X \leq 7; 8) = 0.0003 + \ldots + 0.1396 = 0.4529$

30.
a = 12, b = 36, n = 6, X = 3

30. continued

$$P(A) = \frac{_{12}C_3 \cdot _{36}C_3}{_{48}C_6} = \frac{\frac{12!}{9!\,3!} \cdot \frac{36!}{33!\,3!}}{\frac{48!}{42!\,6!}} = 0.128$$

31.

a. $\frac{_6C_3 \cdot _8C_1}{_{14}C_4} = \frac{\frac{6!}{3!\,3!} \cdot \frac{8!}{7!\,1!}}{\frac{14!}{10!\,4!}} = 0.16$

b. $\frac{_6C_2 \cdot _8C_2}{_{14}C_4} = \frac{\frac{6!}{4!\,2!} \cdot \frac{8!}{6!\,2!}}{\frac{14!}{10!\,4!}} = 0.42$

c. $\frac{_6C_0 \cdot _8C_4}{_{14}C_4} = \frac{\frac{6!}{6!\,0!} \cdot \frac{8!}{4!\,4!}}{\frac{14!}{10!\,4!}} = 0.07$

Graphs are not to scale and are intended to convey a general idea. Answers are generated using Table E. Answers generated using the TI-83 will vary slightly. Some TI-83 answers are shown.

EXERCISE SET 6-1

1.
The characteristics of the normal distribution are:
1. It is bell-shaped.
2. It is symmetric about the mean.
3. The mean, median, and mode are equal.
4. It is continuous.
5. It never touches the X-axis.
6. The area under the curve is equal to one.
7. It is unimodal.

3.
One or 100%.

5.
68%, 95%, 99.7%

7.
The area is found by looking up $z = 0.75$ in Table E and subtracting 0.5.
Area $= 0.7734 - 0.5 = 0.2734$

0   0.75

9.
The area is found by looking up $z = -2.07$ in Table E and subtracting from 0.5.
Area $= 0.5 - 0.0192 = 0.4808$

-2.07 0

11.
The area is found by looking up $z = 0.29$ in Table E and subtracting it from 1.
Area $= 1 - 0.6141 = 0.3859$

11. continued

0   0.29

13.
The area is found by looking up $z = -1.39$ in Table E.
Area $= 0.0823$

-1.39 0

15.
The area is found by looking up the values 1.05 and 1.78 in Table E and subtracting the areas.
Area $= 0.9625 - 0.8531 = 0.1094$

0   1.05   1.78

17.
The area is found by looking up the values -1.56 and -1.83 in Table E and subtracting the areas.
Area $= 0.0594 - 0.0336 = 0.0258$

-1.83 -1.56 0

19.
The area is found by looking up the values
-1.46 and -1.98 in Table E and
subtracting the areas.
Area $= 0.0721 - 0.0239 = 0.0482$

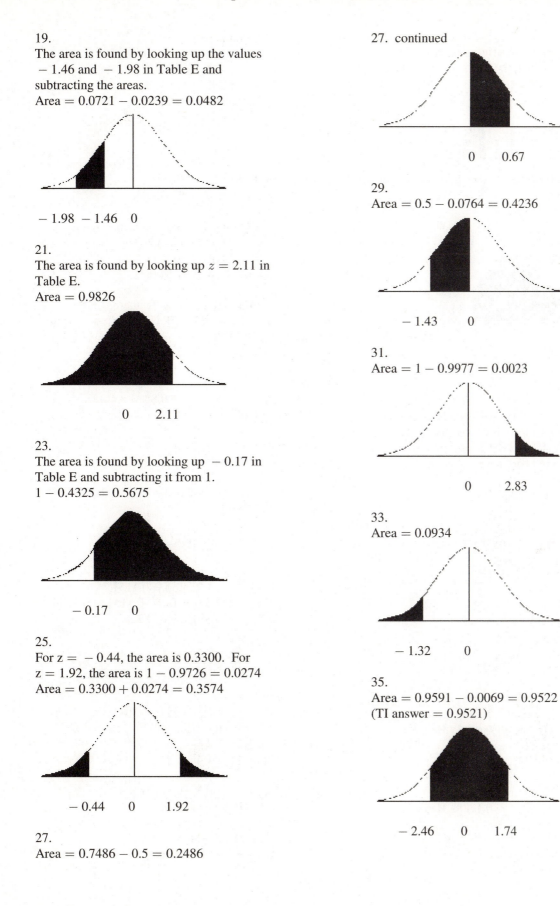

-1.98 -1.46 $\ 0$

21.
The area is found by looking up $z = 2.11$ in
Table E.
Area $= 0.9826$

$0 \quad 2.11$

23.
The area is found by looking up -0.17 in
Table E and subtracting it from 1.
$1 - 0.4325 = 0.5675$

$-0.17 \quad 0$

25.
For $z = -0.44$, the area is 0.3300. For
$z = 1.92$, the area is $1 - 0.9726 = 0.0274$
Area $= 0.3300 + 0.0274 = 0.3574$

$-0.44 \quad 0 \quad 1.92$

27.
Area $= 0.7486 - 0.5 = 0.2486$

27. continued

$0 \quad 0.67$

29.
Area $= 0.5 - 0.0764 = 0.4236$

$-1.43 \quad 0$

31.
Area $= 1 - 0.9977 = 0.0023$

$0 \quad 2.83$

33.
Area $= 0.0934$

$-1.32 \quad 0$

35.
Area $= 0.9591 - 0.0069 = 0.9522$
(TI answer $= 0.9521$)

$-2.46 \quad 0 \quad 1.74$

37.
Area $= 0.9985 - 0.9279 = 0.0706$
(TI answer $= 0.0707$)

0   1.46 2.97

39.
Area $= 0.9222$

0   1.42

41.
Since the z score is on the left side of 0, use the negative z table. Areas in the negative z table are in the tail, so we will use $0.5 - 0.4175 = 0.0825$ as the area. The closest z score corresponding to an area of 0.0825 is $z = -1.39$.
(TI answer $= -1.3885$)

43.
$z = -2.08$, found by using the negative z table.
(TI answer $= -2.0792$)

45.
Use the negative z table and $1 - 0.8962 = 0.1038$ for the area. The z score is $z = -1.26$.
(TI answer $= -1.2602$)

47.
a. Using the negative z table, area $= 1 - 0.9887 = 0.0113$. Hence $z = -2.28$.
(TI answer $= -2.2801$)

b. Using the negative z table, area $= 1 - 0.8212 = 0.1788$. Hence $z = -0.92$.
(TI answer $= -0.91995$)

47. continued
c. Using the negative z table, area $= 1 - 0.6064 = 0.3936$. Hence $z = -0.27$.
(TI answer $= -0.26995$)

49.
a. For total area $= 0.05$, there will be area $= 0.025$ in each tail. The z scores are ± 1.96.
(TI answer $= \pm 1.95996$)

b. For total area $= 0.10$, there will be area $= 0.05$ in each tail. The z scores are $z = \pm 1.645$.
(TI answer $= \pm 1.64485$)

c. For total area $= 0.01$, there will be area $= 0.005$ in each tail. The z scores are $z = \pm 2.58$.
(TI answer $= \pm 2.57583$)

51.
$P(-1 < z < 1) = 0.8413 - 0.1587$
$= 0.6826$

$P(-2 < z < 2) = 0.9772 - 0.0228$
$= 0.9544$

$P(-3 < z < 3) = 0.9987 - 0.0013$
$= 0.9974$

They are very close.

53.
For $z = -1.2$, area $= 0.1151$
Area (left side) $= 0.5 - 0.1151 = 0.3849$
$0.8671 - 0.3849 = 0.4822$
Area (right side) $= 0.4822 + 0.5 = 0.9822$
For area $= 0.9822$, $z = 2.10$
Thus, $P(-1.2 < z < 2.10) = 0.8671$

55.
For $z = -0.5$, area $= 0.3085$
$0.3085 - 0.2345 = 0.074$
For area $= 0.074$, $z = -1.45$
Thus, $P(-1.45 < z < -0.5) = 0.2345$

For $z = -0.5$, area $= 0.3085$
$0.5 - 0.3085 = 0.1915$
$0.2345 - 0.1915 = 0.043$
$0.5 + 0.043 = 0.543$
For area $= 0.543$, $z = 0.11$
Thus, $P(-0.5 < z < 0.11) = 0.2345$

57.

$$y = \frac{e^{\frac{-(X-0)^2}{2(1)^2}}}{1\sqrt{2\pi}} = \frac{e^{\frac{-X^2}{2}}}{\sqrt{2\pi}}$$

EXERCISE SET 6-2

1.
$z = \frac{\$3.50 - \$5.81}{\$0.81} = -2.85$

$P(z < -2.85) = 0.0022$ or 0.22%

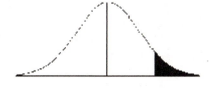

$-2.85 \qquad 0$

3.
a. $z = \frac{750,000 - 706,242}{52,145} = 0.84$

$P(z > 0.84) = 1 - 0.7995 = 0.2005$ or
20.05%
(TI answer = 0.2007)

$0 \qquad 0.84$

b. $z = \frac{600,000 - 706,242}{52,145} = -2.04$

$z = \frac{700,000 - 706,242}{52,145} = -0.12$

$P(-2.04 < z < -0.12) = 0.4522 - 0.0207$
$P = 0.4315$ or 43.15%
(TI answer = 0.4316)

$-2.04 \quad -0.12$

5.
a. $z = \frac{200 - 225}{10} = -2.5$

$z = \frac{220 - 225}{10} = -0.5$

5. continued
$P(-2.5 < z < -0.5) =$
$0.3085 - 0.0062 = 0.3023$ or 30.23%

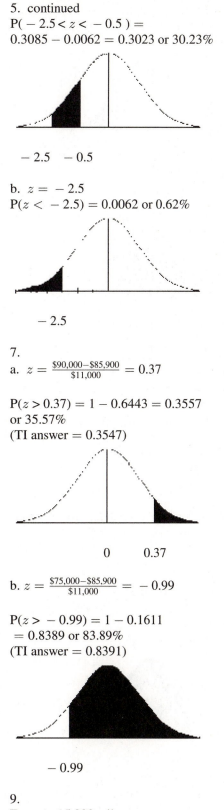

$-2.5 \quad -0.5$

b. $z = -2.5$
$P(z < -2.5) = 0.0062$ or 0.62%

-2.5

7.
a. $z = \frac{\$90,000 - \$85,900}{\$11,000} = 0.37$

$P(z > 0.37) = 1 - 0.6443 = 0.3557$
or 35.57%
(TI answer = 0.3547)

$0 \qquad 0.37$

b. $z = \frac{\$75,000 - \$85,900}{\$11,000} = -0.99$

$P(z > -0.99) = 1 - 0.1611$
$= 0.8389$ or 83.89%
(TI answer = 0.8391)

-0.99

9.
For $x \geq 15,000$ miles:

$z = \frac{15,000 - 12,494}{1290} = 1.94$

9. continued

$P(z > 1.94) = 1 - 0.9738 = 0.0262$
(TI answer = 0.02603)

0   1.94

For $x < 8000$ miles:

$z = \frac{8000 - 12,494}{1290} = -3.48$

$P(z < -3.48) = 0.0003$
(TI answer = 0.00025)

−3.48   0

For $x < 6000$ miles:

$z = \frac{6000 - 12,494}{1290} = -5.03$

$P(z < -5.03) = 0.0001$

Maybe it would be good to know why it had only been driven less than 6000 miles.

11.
a. $z = \frac{1000 - 3262}{1100} = -2.06$

$P(z \geq -2.06) = 1 - 0.0197 = 0.9803$ or 98.03%
(TI answer = 0.9801)

−2.06

b. $z = \frac{4000 - 3262}{1100} = 0.67$

$P(z > 0.67) = 1 - 0.7486 = 0.2514$ or 25.14%
(TI answer = 0.2511)

11b. continued

0   0.67

c. $z = \frac{3000 - 3262}{1100} = -0.24$

$P(-0.24 < z < 0.67) =$
$0.7486 - 0.4052 = 0.3434$ or 34.34%
(TI answer = 0..3430)

−0.24   0.67

13.
a. $z = \frac{4 - 10.3}{2.7} = -2.33$

$z = \frac{9 - 10.3}{2.7} = -0.48$

$P(-2.33 < z < -0.48) =$
$0.3156 - 0.0099 = 0.3057$
(TI answer = 0.30527)

−2.33 −0.48   0

b. $z = \frac{5 - 10.3}{2.7} = -1.96$

$z = \frac{10 - 10.3}{2.7} = -0.11$

$P(z < -1.96 \text{ or } z > -0.11) =$
$0.0250 + (1 - 0.4562) = 0.5688$
(TI answer = 0.56906)

−1.96 −0.11

13. continued
c. If the customer knew the mean and standard deviation, the approximate wait time would be between the mean plus 2 standard deviations and the mean minus 2 standard deviations.

15.
a. $z = \frac{15-23.5}{3.6} = -2.36$

$z = \frac{22-23.5}{3.6} = -0.42$

$P(-2.36 < z < -0.42) =$
$0.3372 - 0.0091 = 0.3281$

$-2.36 \quad -0.42$

b. $z = \frac{18-23.5}{3.6} = -1.53$

$z = \frac{25-23.5}{3.6} = 0.42$

$P(z < -1.53 \text{ or } z > 0.42) =$
$0.0630 + (1 - 0.6628) =$
$0.063 + 0.3372 = 0.4002$

$-1.53 \quad 0 \quad 0.42$

c. For 15 minutes, $z = -2.36$.
$P(z < -2.36) = 0.0091$
Since the probability is small, it is not likely that a person would be seated in less than 15 minutes.

17.
$z = \frac{63-71.2}{3.4} = -2.41$

$P(z < -2.41) = -0.008 \text{ or } 0.8\%$
(TI answer = 0.00794)

17. continued

$-2.41 \qquad 0$

A temperature of 63° is unlikely since the probability is about 0.8%.

19.
The middle 80% means that 40% of the area will be on either side of the mean. The corresponding z scores will be ± 1.28.
$x = -1.28(92) + 1810 = 1692.24$ sq. ft.
$x = 1.28(92) + 1810 = 1927.76$ sq. ft.
(TI answers: 1927.90 maximum, 1692.10 minimum)

1692   1810   1928

21.
$z = \frac{1200-949}{100} = 2.51$

$P(z > 2.51) = 1 - 0.9940 = 0.006 \text{ or } 0.6\%$

0   2.51

For the least expensive 10%, the area is 0.4 on the left side of the curve. Thus,
$z = -1.28$.
$x = -1.28(100) + 949 = \$821$

23.
The middle 60% means that 30% of the area will be on either side of the mean. The corresponding z scores will be ± 0.84.
$x = -0.84(1150) + 8256 = \$7290$
$x = 0.84(1150) + 8256 = \$9222$
(TI answer: $\$7288.14 < \mu < \$9223.86$)

23. continued

$7290 $8256 $9222

25.
For the longest 10%, the area is 0.90. Thus,
$z = 1.28$
Since $\sigma^2 = 2.1$, $\sigma = \sqrt{2.1} = 1.449$
$x = 1.28(1.449) + 4.8$
$x = 6.65$ or 6.7 days
(TI answer = 6.657)
For the shortest 30%, the area is 0.30. Thus,
$z = -0.52$.
$x = -0.52(1.449) + 4.8$
$x = 4.047$ days or 4.05 days
(TI answer = 4.040)

27.
The bottom 18% area is 0.18. Thus,
$z = -0.92$.
$x = -0.92(6256) + 24{,}596 = \$18{,}840.48$
(TI answer = $18,869.48)

29.
The 10% to be exchanged would be at the left, or bottom, of the curve; therefore, area = 0.10. The corresponding z score will be -1.28.
$x = -1.28(5) + 25 = 18.6$ months.

18.6   25

31.
a. $\mu = 120$ $\sigma = 20$
b. $\mu = 15$ $\sigma = 2.5$
c. $\mu = 30$ $\sigma = 5$

33.
There are several mathematical tests that can be used including drawing a histogram and calculating Pearson's index of skewness.

35.
5.48% area in the right tail of the curve means that area = 0.9452. Thus, $z = 1.60$.

Since $z = \frac{X - \mu}{\sigma}$:
$1.60 = \frac{110 - 105}{\sigma}$
$1.60\sigma = 5$
$\sigma = 3.125$

37.
1.25% of the area in each tail means that area = 0.0125 in the left tail and area = 0.9875 in the right tail. Thus, $z = \pm 2.24$.
Then $\mu = \frac{42 + 48}{2} = 45$ and $x = \mu + z\sigma$.
$48 = 45 + 2.24\sigma$
$\sigma = 1.34$

39.
Histogram:

The histogram shows a positive skew.

$PI = \frac{3(970.2 - 853.5)}{376.5} = 0.93$

$IQR = Q_3 - Q_1 = 910 - 815 = 95$
$1.5(IQR) = 1.5(95) = 142.5$
$Q_1 - 142.5 = 672.5$
$Q_3 + 142.5 = 1052.5$
There are several outliers.
Conclusion: The distribution is not normal.

41.
Histogram:

The histogram shows a positive skew.

$PI = \frac{3(90 - 59)}{89.598} = 1.04$

$IQR = Q_3 - Q_1 = 111 - 32 = 79$

41. continued

$1.5(\text{IQR}) = 1.5(79) = 118.5$

$Q_1 - 118.5 = -86.5$

$Q_3 + 118.5 = 229.5$

There are two outliers.

Conclusion: The distribution is not normal.

EXERCISE SET 6-3

1.

The distribution is called the sampling distribution of sample means.

3.

The mean of the sample means is equal to the population mean.

5.

The distribution will be approximately normal when sample size is large.

7.

$z = \frac{\overline{X} - \mu}{\sigma / \sqrt{n}}$

9.

a. $z = \frac{\overline{X} - \mu}{\frac{\sigma}{\sqrt{n}}} = \frac{\$25,000 - \$26,489}{\frac{\$3204}{\sqrt{36}}} = -2.79$

$P(z < -2.79) = 0.0026$ or 0.26%

(TI answer $= 0.0026$)

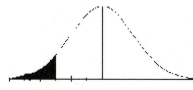

$\$25,000 \quad \$26,489$

b. $z = \frac{\overline{X} - \mu}{\frac{\sigma}{\sqrt{n}}} = \frac{\$26,000 - \$26,489}{\frac{\$3204}{\sqrt{36}}} = -0.92$

$P(z > -0.92) = 1 - 0.1788 = 0.8212$

(TI answer $= 0.8201$)

$26,000 \quad 26,489$

c. $z = \frac{\overline{X} - \mu}{\frac{\sigma}{\sqrt{n}}} = \frac{\$24,000 - \$26,489}{\frac{\$3204}{\sqrt{36}}} = -4.66$

9c. continued

$z = \frac{\overline{X} - \mu}{\frac{\sigma}{\sqrt{n}}} = \frac{\$26,000 - \$26,489}{\frac{\$3204}{\sqrt{36}}} = -0.92$

$P(-4.66 < z < -0.92) =$

$0.1788 - 0.0001 = 0.1787$

(TI answer $= 0.1799$)

$\$24,000 \quad \$26,000$

11.

$z = \frac{\overline{X} - \mu}{\frac{\sigma}{\sqrt{n}}} = \frac{220 - 220}{\frac{16.3}{\sqrt{35}}} = 0$

$z = \frac{\overline{X} - \mu}{\frac{\sigma}{\sqrt{n}}} = \frac{222 - 220}{\frac{16.3}{\sqrt{35}}} = 0.73$

$P(0 < z < 0.73) = 0.7673 - 0.5 = 0.2673$ or 26.73%

(TI answer $= 0.2660$)

$220 \quad 222$

13.

For $x < 20$ mpg:

$z = \frac{\overline{X} - \mu}{\frac{\sigma}{\sqrt{n}}} = \frac{20 - 21}{\frac{2.9}{\sqrt{25}}} = -1.72$

$P(z < -1.72) = 0427$ or 4.27%

(TI answer $= 0.0423$)

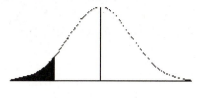

$20 \quad 21$

For $20 < x < 25$:

$z = \frac{\overline{X} - \mu}{\frac{\sigma}{\sqrt{n}}} = \frac{25 - 21}{\frac{2.9}{\sqrt{25}}} = 6.90$

13. continued

$$z = \frac{\overline{X} - \mu}{\frac{\sigma}{\sqrt{n}}} = \frac{20 - 21}{\frac{2.9}{\sqrt{25}}} = -1.72$$

$P(20 < z < 25) = 0.9999 - 0.0427$
$= 0.9572$ or 95.72%
(TI answer = 0.9577)

20   21   25

15.

a. $z = \frac{\overline{X} - \mu}{\sigma} = \frac{670 - 660}{35} = 0.29$

$P(z > 0.29) = 1 - 0.6141 = 0.3859$
(TI answer = 0.3875)

0   0.29

b. $z = \frac{\overline{X} - \mu}{\frac{\sigma}{\sqrt{n}}} = \frac{670 - 660}{\frac{35}{\sqrt{10}}} = 0.90$

$P(z > 0.90) = 1 - 0.8159 = 0.1841$
(TI answer = 0.1831)

660   670

c. Individual values are more variable than means.

17.

$$z = \frac{\overline{X} - \mu}{\frac{\sigma}{\sqrt{n}}} = \frac{120 - 123}{\frac{21}{\sqrt{15}}} = -0.55$$

$$z = \frac{\overline{X} - \mu}{\frac{\sigma}{\sqrt{n}}} = \frac{126 - 123}{\frac{21}{\sqrt{15}}} = 0.55$$

$P(-0.55 < z < 0.55) = 0.7088 - 0.2912$
$= 0.4176$ or 41.76%
(TI answer = 0.4199)

17. continued

120   123   126

19.

$$z = \frac{\overline{X} - \mu}{\frac{\sigma}{\sqrt{n}}} = \frac{1980 - 2000}{\frac{187.5}{\sqrt{50}}} = -0.75$$

$$z = \frac{\overline{X} - \mu}{\frac{\sigma}{\sqrt{n}}} = \frac{1990 - 2000}{\frac{187.5}{\sqrt{50}}} = -0.38$$

$P(-0.75 < z < -0.38) =$
$0.3520 - 0.2266 = 0.1254$
(TI answer = 0.12769)

1980 1990 2000

21.

a. $z = \frac{X - \mu}{\sigma} = \frac{30 - 30.85}{3.6} = -0.24$

$P(z < -0.24) = 0.4052$ or 40.52%

−0.24   0

b. $z = \frac{30 - 30.85}{\frac{3.6}{\sqrt{32}}} = -1.34$

$P(z < -1.34) = 0.0901$ or 9.01%

30   30.85

c. Yes, since the probability is slightly more than 40%.

21. continued
d. It's possible since the probability is about 9%.

23.
a. $z = \frac{220-215}{15} = 0.33$

$P(z > 0.33) = 1 - 0.6293 = 0.3707$ or 37.07%
(TI answer = 0.3694)

0   0.33

b. $z = \frac{220-215}{\frac{15}{\sqrt{25}}} = 1.67$

$P(z > 1.67) = 1 - 0.9525 = 0.0475$ or 4.75%
(TI answer = 0.04779)

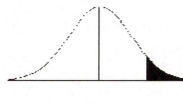

215   220

25.
$z = \frac{800-777}{\frac{77}{\sqrt{50}}} = 2.11$

$P(z > 2.11) = 1 - 0.9826 = 0.0174$

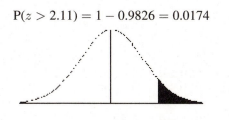

0   2.11

No, the central limit theorem applies.

27.
Since $50 > 0.05(800)$ or 40, the correction factor is necessary.
It is $\sqrt{\frac{800-50}{800-1}} = 0.969$

$z = \frac{\overline{X}-\mu}{\frac{\sigma}{\sqrt{n}} \cdot \sqrt{\frac{N-n}{n-1}}} = \frac{83,500-82,000}{\frac{5000}{\sqrt{50}}(0.969)} = 2.19$

27. continued
$P(z > 2.19) = 1 - 0.9857 = 0.0143$ or 1.43%

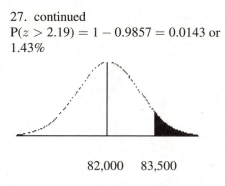

82,000   83,500

29.
$\sigma_x = \frac{\sigma}{\sqrt{n}} = \frac{15}{\sqrt{100}} = 1.5$
$2(1.5) = \frac{15}{\sqrt{n}}$
$3 \cdot \sqrt{n} = 15$
$\sqrt{n} = 5$
$n = 25$, the sample size necessary to double the standard error.

EXERCISE SET 6-4

1.
When p is approximately 0.5, and as n increases, the shape of the binomial distribution becomes similar to the normal distribution. The normal approximation should be used only when $n \cdot p$ and $n \cdot q$ are both greater than or equal to 5. The correction for continuity is necessary because the normal distribution is continuous and the binomial is discrete.

2.
For each problem use the following formulas:
$\mu = np$ $\sigma = \sqrt{npq}$ $z = \frac{\overline{X}-\mu}{\sigma}$
Be sure to correct each X for continuity.

a. $\mu = 0.5(30) = 15$
$\sigma = \sqrt{(0.5)(0.5)(30)} = 2.74$

$z = \frac{17.5-15}{2.74} = 0.91$ area = 0.8186

$z = \frac{18.5-15}{2.74} = 1.28$ area = 0.8997

$P(17.5 < X < 18.5) = 0.8997 - 0.8186$
$= 0.0811 = 8.11\%$

2a. continued

15   17.5 18.5

b. $\mu = 0.8(50) = 40$
$\sigma = \sqrt{(50)(0.8)(0.2)} = 2.83$

$z = \frac{43.5-40}{2.83} = 1.24$ area $= 0.8925$

$z = \frac{44.5-40}{2.83} = 1.59$ area $= 0.9441$

$P(43.5 < X < 44.5) = 0.9441 - 0.8925$
$= 0.0516$ or 5.16%

40   43.5 44.5

c. $\mu = 0.1(100) = 10$
$\sigma = \sqrt{(0.1)(0.9)(100)} = 3$

$z = \frac{11.5-10}{3} = 0.50$ area $= 0.6915$

$z = \frac{12.5-10}{3} = 0.83$ area $= 0.7967$

$P(11.5 < X < 12.5) = 0.7967 - 0.6915$
$= 0.1052$ or 10.52%

10   11.5 12.5

d. $\mu = 10(0.5) = 5$
$\sigma = \sqrt{(0.5)(0.5)(10)} = 1.58$

$z = \frac{6.5-5}{1.58} = 0.95$ area $= 0.8289$

$P(X \geq 6.5) = 1 - 0.8289 = 0.1711$ or
17.11%

2d. continued

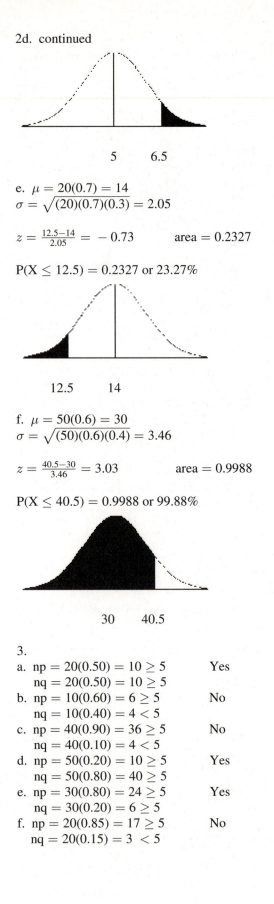

5   6.5

e. $\mu = 20(0.7) = 14$
$\sigma = \sqrt{(20)(0.7)(0.3)} = 2.05$

$z = \frac{12.5-14}{2.05} = -0.73$ area $= 0.2327$

$P(X \leq 12.5) = 0.2327$ or 23.27%

12.5   14

f. $\mu = 50(0.6) = 30$
$\sigma = \sqrt{(50)(0.6)(0.4)} = 3.46$

$z = \frac{40.5-30}{3.46} = 3.03$ area $= 0.9988$

$P(X \leq 40.5) = 0.9988$ or 99.88%

30   40.5

3.
a. np $= 20(0.50) = 10 \geq 5$ Yes
    nq $= 20(0.50) = 10 \geq 5$
b. np $= 10(0.60) = 6 \geq 5$ No
    nq $= 10(0.40) = 4 < 5$
c. np $= 40(0.90) = 36 \geq 5$ No
    nq $= 40(0.10) = 4 < 5$
d. np $= 50(0.20) = 10 \geq 5$ Yes
    nq $= 50(0.80) = 40 \geq 5$
e. np $= 30(0.80) = 24 \geq 5$ Yes
    nq $= 30(0.20) = 6 \geq 5$
f. np $= 20(0.85) = 17 \geq 5$ No
    nq $= 20(0.15) = 3 < 5$

5.

$p = \frac{2}{5} = 0.4 \qquad \mu = 400(0.4) = 160$

$\sigma = \sqrt{(400)(0.4)(0.6)} = 9.8$

$z = \frac{170.5 - 160}{9.8} = 1.07$

$P(X \leq 170.5) = 0.8577$ or 85.77%

160   170.5

7.

$\mu = 400(0.3) = 120$

$\sigma = \sqrt{(400)(0.3)(0.7)} = 9.17$

$z = \frac{99.5 - 120}{9.17} = -2.24$

$P(X \geq 99.5) = 1 - 0.0125 = 0.9875$
(TI answer = 0.9873)

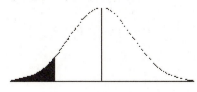

99.5   120

9.

$\mu = 200(0.261) = 52.2$

$\sigma = \sqrt{(200)(0.261)(0.739)} = 6.21$

$z = \frac{50.5 - 52.2}{6.21} = -0.27$

$P(X \leq 50.5) = 0.3936$

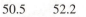

50.5   52.2

11.

$\mu = 300(0.803) = 240.9$

$\sigma = \sqrt{(300)(0.803)(0.197)} = 6.89$

$X < \frac{3}{4}(300)$ or $X < 225$

$z = \frac{224.5 - 240.9}{6.89} = -2.38$

11. continued

$P(X < 224.5) = 0.0087$

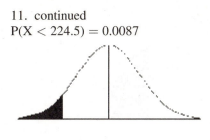

224.5   240.9

13.

$\mu = 350(0.35) = 122.5$

$\sigma = \sqrt{(350)(0.35)(0.65)} = 8.92$

$z = \frac{99.5 - 122.5}{8.92} = -2.58 \qquad$ area $= 0.0049$

$P(X > 99.5) = 1 - 0.0049 = 0.9951$ or
99.51% (TI answer = 0.9950)
Yes; it is likely that 100 or more people
would favor the parking lot.

99.5   122.5

15.

a. $n(0.1) = 5 \qquad\qquad n \geq 50$
b. $n(0.3) = 5 \qquad\qquad n \geq 17$
c. $n(0.5) = 5 \qquad\qquad n \geq 10$
d. $n(0.2) = 5 \qquad\qquad n \geq 25$
e. $n(0.1) = 5 \qquad\qquad n \geq 50$

REVIEW EXERCISES - CHAPTER 6

1.

a. $0.9744 - 0.5 = 0.4744$

0   1.95

1. continued
b. 0.6443 − 0.5 − 0.1443

0   0.37

c. 0.9656 − 0.9066 = 0.0590

0   1.32 1.82

d. 0.9798 − 0.1469 = 0.8329
(TI answer = 0.83296)

− 1.05   0   2.05

e. 0.7019 − 0.4880 = 0.2139

− 0.03   0   0.53

f. 0.8643 − 0.0359 = 0.8284

− 1.80   0   1.10

g. 1 − 0.9767 = 0.0233

1g. continued

0   1.99

h. 1 − 0.0869 = 0.9131

− 1.36   0

i. 0.0183

− 2.09   0

j. 0.9535

0   1.68

3.
a. $z = \frac{\$6000 - \$5274}{\$600} = 1.21$

$P(z > 1.21) = 1 - 0.8869 = 0.1131$

0   1.21

b. For the middle 50%, 25% of the area is on each side of 0. Thus, $z = \pm 0.67$

3b. continued
$x = 0.67(600) + 5274 = \$5676$
$x = -0.67(600) + 5274 = \$4872$
(TI answers: \$4869.31 to \$5678.69)

5.
a. $z = \frac{476-476}{22} = 0$

$z = \frac{500-476}{22} = 1.09$

$P(0 < z < 1.09) = 0.8621 - 0.5 = 0.3621$
or 36.21%

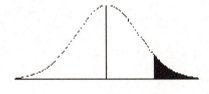

0   1.09

b. $z = \frac{450-476}{22} = -1.18$

$P(z < -1.18) = 0.1190$ or 11.9%

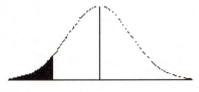

−1.18   0

c. $z = \frac{510-476}{22} = 1.55$

$P(z > 1.55) = 1 - 0.9394 = 0.0606$ or
6.06%

0   1.55

7.
For 15% costs, area = 0.85
$z = 1.04$
$X = 1.04(10.50) + 120 = \$130.92$

9.
Histogram:

9. continued

The histogram shows a positive skew.

$PI = \frac{3(2136.1-1755)}{1171.7} = 0.98$

$IQR = Q_3 - Q_1$
$IQR = 2827 - 1320 = 1507$
$1.5(IQR) = 1.5(1507) = 2260.5$
$Q_1 - 2260.5 = -940.5$
$Q_3 + 2260.5 = 5087.5$
There are no outliers.
Conclusion: The distribution is not normal.

11.
a. $z = \frac{X-\mu}{\frac{\sigma}{\sqrt{n}}} = \frac{27 - 25.7}{\frac{3.75}{\sqrt{40}}} = 2.19$

$P(\overline{X} > 27) = 1 - 0.9857 = 0.0143$
(TI answer = 0.0142)

25.7   27

b. $z = \frac{X-\mu}{\frac{\sigma}{\sqrt{n}}} = \frac{\$60 - \$61.50}{\frac{\$5.89}{\sqrt{50}}} = -1.80$

$P(\overline{X} > 60) = 1 - 0.0359 = 0.9641$

\$60   \$61.50

13.
$\mu = 120(0.173) = 20.76$
$\sigma = \sqrt{120(0.173)(0.827)} = 4.14$

$z = \frac{20.5 - 20.76}{4.14} = -0.06$

$z = \frac{34.5 - 20.76}{4.14} = 3.32$

13. continued
$P(20.5 < X < 34.5) = 0.9995 - 0.4761$
$= 0.5234$
(TI answer = 0.52456)

20.5   20.76   34.5

15.
For fewer than 10 holding multiple jobs:
$\mu = 150(0.053) = 7.95$
$\sigma = \sqrt{(150)(0.053)(0.947)} = 2.744$

$z = \frac{9.5 - 7.95}{2.74} = 0.56$

$P(X < 9.5) = 0.7123$
(TI answer = 0.7139)

7.95   9.5

For more than 50 not holding multiple jobs:
$\mu = 150(0.947) = 142.05$
$\sigma = \sqrt{150(0.947)(0.053)} = 2.744$

$z = \frac{50.5 - 142.05}{2.744} = -33.37$

$P(X > 50.5) = 1 - 0.0001 = 0.9999$
(TI answer = 0.9999)

50.5   142.05

17.
$\mu = 200(0.2) = 40$
$\sigma = \sqrt{(200)(0.2)(0.8)} = 5.66$

$z = \frac{49.5 - 40}{5.66} = 1.68$

17. continued
$P(X \geq 49.5) = 1 - 0.9535 = 0.0465$ or
4.65%

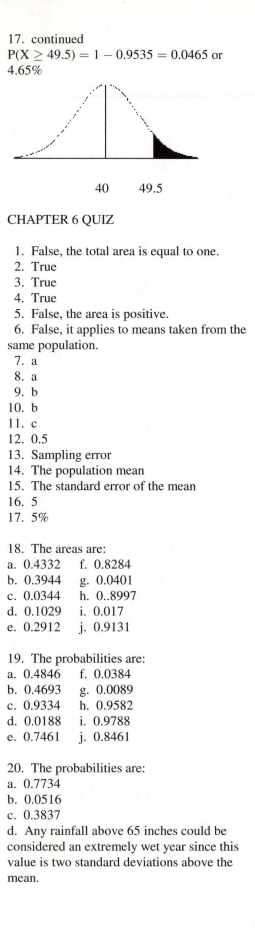

40   49.5

CHAPTER 6 QUIZ

1. False, the total area is equal to one.
2. True
3. True
4. True
5. False, the area is positive.
6. False, it applies to means taken from the same population.
7. a
8. a
9. b
10. b
11. c
12. 0.5
13. Sampling error
14. The population mean
15. The standard error of the mean
16. 5
17. 5%

18. The areas are:
a. 0.4332   f. 0.8284
b. 0.3944   g. 0.0401
c. 0.0344   h. 0..8997
d. 0.1029   i. 0.017
e. 0.2912   j. 0.9131

19. The probabilities are:
a. 0.4846   f. 0.0384
b. 0.4693   g. 0.0089
c. 0.9334   h. 0.9582
d. 0.0188   i. 0.9788
e. 0.7461   j. 0.8461

20. The probabilities are:
a. 0.7734
b. 0.0516
c. 0.3837
d. Any rainfall above 65 inches could be considered an extremely wet year since this value is two standard deviations above the mean.

21. The probabilities are:
a. 0.0668   c. 0.4649
b. 0.0228   d. 0.0934

22. The probabilities are:
a. 0.4525   c. 0.3707
b. 0.3707   d. 0.019

23. The probabilities are:
a. 0.0013   c. 0.0081
b. 0.5   d. 0.5511

24. The probabilities are:
a. 0.0037   c. 0.5
b. 0.0228   d. 0.3232

25. 8.804 cm
26. The lowest acceptable score is 121.24.
27. 0.015
28. 0.9738
29. 0.0495; no
30. 0.0455
31. 0.8577
32. 0.0495
33. The distribution is not normal.
34. The distribution is approximately normal.

Note: Answers may vary due to rounding.

EXERCISE SET 7-1

1.
A point estimate of a parameter specifies a specific value such as $\mu = 87$, whereas an interval estimate specifies a range of values for the parameter such as $84 < \mu < 90$. The advantage of an interval estimate is that a specific confidence level (say 95%) can be selected, and one can be 95% confident that the parameter being estimated lies in the interval.

3.
The margin of error is the likely range of values above or below the statistic that may contain the parameter.

5.
A good estimator should be unbiased, consistent, and relatively efficient.

7.
To determine sample size, the maximum error of estimate and the degree of confidence must be specified and the population standard deviation must be known.

9.
a. 2.58   d. 1.65
b. 2.33   e. 1.88
c. 1.96

11.
a. $\overline{X} = 16.6$ is the point estimate for μ.

b. $\overline{X} - z_{\frac{\alpha}{2}}\left(\frac{s}{\sqrt{n}}\right) < \mu < \overline{X} + z_{\frac{\alpha}{2}}\left(\frac{s}{\sqrt{n}}\right)$
$16.6 - (1.96)\left(\frac{2.8}{\sqrt{35}}\right) < \mu <$
$\qquad 16.6 + (1.96)\left(\frac{15}{\sqrt{35}}\right)$
$16.6 - 0.93 < \mu < 16.6 + 0.93$
$15.67 < \mu < 17.53$ or $15.7 < \mu < 17.5$

c. $16.6 - (2.58)\left(\frac{2.8}{\sqrt{35}}\right) < \mu <$
$\qquad 16.6 + (2.58)\left(\frac{2.8}{\sqrt{35}}\right)$
$16.6 - 1.22 < \mu < 16.6 + 1.22$
$15.38 < \mu < 17.82$ or $15.4 < \mu < 17.8$

d. The 99% confidence interval is larger because the confidence level is larger.

13.
$\overline{X} - z_{\frac{\alpha}{2}}\left(\frac{\sigma}{\sqrt{n}}\right) < \mu < \overline{X} + z_{\frac{\alpha}{2}}\left(\frac{\sigma}{\sqrt{n}}\right)$
$1.8 - 1.65\left(\frac{0.33}{\sqrt{50}}\right) < \mu < 1.8 + 1.65\left(\frac{0.33}{\sqrt{50}}\right)$
$1.8 - 0.08 < \mu < 1.8 + 0.08$
$1.72 < \mu < 1.88$ The estimate is lower.

15.
$\overline{X} - z_{\frac{\alpha}{2}}\left(\frac{s}{\sqrt{n}}\right) < \mu < \overline{X} + z_{\frac{\alpha}{2}}\left(\frac{s}{\sqrt{n}}\right)$
$\$150{,}000 - 1.96\left(\frac{15{,}000}{\sqrt{35}}\right) < \mu <$
$\qquad \$150{,}000 + 1.96\left(\frac{15{,}000}{\sqrt{35}}\right)$
$\$150{,}000 - 4969.51 < \mu <$
$\qquad \$150{,}000 + 4969.51$

$\$145{,}030 < \mu < \$154{,}970$
(TI answer: $\$145{,}031 < \mu < \$154{,}969$)

17.
$\overline{X} = 5000 \qquad s = 900$
$5000 - 1.96\left(\frac{900}{\sqrt{415}}\right) < \mu < 5000 +$
$\qquad 1.96\left(\frac{900}{\sqrt{415}}\right)$
$4913 < \mu < 5087$ 4000 hours does not seem reasonable since it is outside this interval.

19.
$\overline{X} - z_{\frac{\alpha}{2}}\left(\frac{s}{\sqrt{n}}\right) < \mu < \overline{X} + z_{\frac{\alpha}{2}}\left(\frac{s}{\sqrt{n}}\right)$
$61.2 - 1.96\left(\frac{7.9}{\sqrt{84}}\right) < \mu < 61.2 + 1.96\left(\frac{7.9}{\sqrt{84}}\right)$
$61.2 - 1.69 < \mu < 61.2 + 1.69$
$59.5 < \mu < 62.9$

21.
$n = \left[\frac{z_{\frac{\alpha}{2}} \sigma}{E}\right]^2 = \left[\frac{(2.58)(6.43)}{1.5}\right]^2$
$n = (11.0596)^2 = 122.3$ or 123

23.
$n = \left[\frac{z_{\frac{\alpha}{2}} \sigma}{E}\right]^2 = \left[\frac{(1.65)(8)}{2}\right]^2$
$n = (6.6)^2 = 43.56$ or 44

25.
$n = \left[\frac{z_{\frac{\alpha}{2}} \sigma}{E}\right]^2 = \left[\frac{(2.58)(30)}{5}\right]^2$
$n = (15.48)^2 = 239.6$ or 240

EXERCISE SET 7-2

1.
The characteristics of the t-distribution are: It is bell-shaped, symmetrical about the mean, and never touches the x-axis. The

1. continued
mean, median, and mode are equal to 0 and are located at the center of the distribution. The variance is greater than 1. The t-distribution is a family of curves based on degrees of freedom. As sample size increases the t-distribution approaches the normal distribution.

3.
The t-distribution should be used when σ is unknown and n < 30.

4.
a. 2.898 where d. f. = 17
b. 2.074 where d. f. = 22
c. 2.624 where d. f. = 14
d. 1.833 where d. f. = 9
e. 2.093 where d. f. = 19

5.
$\overline{X} - t_{\frac{\alpha}{2}}(\frac{s}{\sqrt{n}}) < \mu < \overline{X} + t_{\frac{\alpha}{2}}(\frac{s}{\sqrt{n}})$
$26.1 - (3.250)(\frac{4.2}{\sqrt{10}}) < \mu <$
$\qquad 26.1 + (3.250)(\frac{4.2}{\sqrt{10}})$
$26.1 - 4.32 < \mu < 26.1 + 4.32$
$21.78 < \mu < 30.42$ or $21.8 < \mu < 30.4$

7.
$\overline{X} = 33.4 \qquad s = 28.7$
$\overline{X} - t_{\frac{\alpha}{2}}(\frac{s}{\sqrt{n}}) < \mu < \overline{X} + t_{\frac{\alpha}{2}}(\frac{s}{\sqrt{n}})$
$33.4 - 1.746(\frac{28.7}{\sqrt{17}}) < \mu < 33.4 + 1.746(\frac{28.7}{\sqrt{17}})$
$33.4 - 12.2 < \mu < 33.4 + 12.2$
$21.2 < \mu < 45.6$

The point estimate is 33.4 and is close to the population mean of 32, which is within the 90% confidence interval. The mean may not be the best estimate since the data value 132 is large and possibly an outlier.

9.
$\overline{X} - t_{\frac{\alpha}{2}}(\frac{s}{\sqrt{n}}) < \mu < \overline{X} + t_{\frac{\alpha}{2}}(\frac{s}{\sqrt{n}})$
$573.8 - 4.032(\frac{46.8}{\sqrt{6}}) < \mu <$
$\qquad 573.8 + 4.032(\frac{46.8}{\sqrt{6}})$
$573.8 - 77.04 < \mu < 573.8 + 77.04$
$496.76 < \mu < 650.84$ or $496.8 < \mu < 650.8$
No, 625 homicides would not be unlikely since it is within the 99% confidence interval.

11.
$\overline{X} - t_{\frac{\alpha}{2}}(\frac{s}{\sqrt{n}}) < \mu < \overline{X} + t_{\frac{\alpha}{2}}(\frac{s}{\sqrt{n}})$
$14.3 - 2.052(\frac{2}{\sqrt{28}}) < \mu < 14.3 + 2.052(\frac{2}{\sqrt{28}})$
$14.3 - 0.8 < \mu < 14.3 + 0.8$
$13.5 < \mu < 15.1$
The employees should allow about 30 minutes.

13.
$\overline{X} - t_{\frac{\alpha}{2}}(\frac{s}{\sqrt{n}}) < \mu < \overline{X} + t_{\frac{\alpha}{2}}(\frac{s}{\sqrt{n}})$
$19.2 - 2.201(\frac{2.1}{\sqrt{12}}) < \mu <$
$\qquad 19.2 + 2.201(\frac{2.1}{\sqrt{12}})$
$19.2 - 1.33 < \mu < 19.2 + 1.33$
$17.87 < \mu < 20.53$
The estimate is higher than the national average.

15.
$\overline{X} - t_{\frac{\alpha}{2}}(\frac{s}{\sqrt{n}}) < \mu < \overline{X} + t_{\frac{\alpha}{2}}(\frac{s}{\sqrt{n}})$
$33.2 - 2.160(\frac{8.3}{\sqrt{14}}) < \mu <$
$\qquad 33.2 + 2.160(\frac{8.3}{\sqrt{14}})$
$33.2 - 4.79 < \mu < 33.2 + 4.79$
$28.41 < \mu < 37.99$ or $28.4 < \mu < 38.0$

17.
$\overline{X} = 51.47 \qquad s = 45.98$
$\overline{X} - t_{\frac{\alpha}{2}}(\frac{s}{\sqrt{n}}) < \mu < \overline{X} + t_{\frac{\alpha}{2}}(\frac{s}{\sqrt{n}})$
$51.47 - 1.746(\frac{45.98}{\sqrt{17}}) < \mu <$
$\qquad 51.47 + 1.746(\frac{45.98}{\sqrt{17}})$
$51.47 - 19.47 < \mu < 51.47 + 19.47$
$32.0 < \mu < 70.9$

19.
Answers will vary.

21.
$\overline{X} = 2.175 \qquad s = 0.585$
For $\mu > \overline{X} - t_{\frac{\alpha}{2}}(\frac{s}{\sqrt{n}})$:
$\mu > 2.175 - 1.729(\frac{0.585}{\sqrt{20}})$
$\mu > 2.175 - 0.226$
Thus, $\mu > \$1.95$ means that one can be 95% confident that the mean revenue is greater than $1.95.

For $\mu < \overline{X} + t_{\frac{\alpha}{2}}(\frac{s}{\sqrt{n}})$:
$\mu < 2.175 + 1.729(\frac{0.585}{\sqrt{20}})$
$\mu < 2.175 + 0.226$

21. continued

Thus, $\mu < \$2.40$ means that one can be 95% confident that the mean revenue is less than $2.40.

EXERCISE SET 7-3

1.

a. $\hat{p} = \frac{40}{80} = 0.5$ $\hat{q} = \frac{40}{80} = 0.5$

b. $\hat{p} = \frac{90}{200} = 0.45$ $\hat{q} = \frac{110}{200} = 0.55$

c. $\hat{p} = \frac{60}{130} = 0.46$ $\hat{q} = \frac{70}{130} = 0.54$

d. $\hat{p} = \frac{35}{60} = 0.58$ $\hat{q} = \frac{25}{60} = 0.42$

e. $\hat{p} = \frac{43}{95} = 0.45$ $\hat{q} = \frac{52}{95} = 0.55$

2.

For each part, change the percent to a decimal by dividing by 100, and find $\hat{q}$ using $\hat{q} = 1 - \hat{p}$.

a. $\hat{p} = 0.25$ $\hat{q} = 1 - 0.25 = 0.75$
b. $\hat{p} = 0.42$ $\hat{q} = 1 - 0.42 = 0.58$
c. $\hat{p} = 0.68$ $\hat{q} = 1 - 0.68 = 0.32$
d. $\hat{p} = 0.55$ $\hat{q} = 1 - 0.55 = 0.45$
e. $\hat{p} = 0.12$ $\hat{q} = 1 - 0.12 = 0.88$

3.

$\hat{p} = 0.39$ $\hat{q} = 0.61$

$\hat{p} - (z_{\frac{\alpha}{2}})\sqrt{\frac{\hat{p}\hat{q}}{n}} < p < \hat{p} + (z_{\frac{\alpha}{2}})\sqrt{\frac{\hat{p}\hat{q}}{n}}$

$0.39 - (1.96)\sqrt{\frac{(0.39)(0.61)}{1500}} < p <$
$\qquad 0.39 + (1.96)\sqrt{\frac{(0.39)(0.61)}{1500}}$

$0.39 - 0.025 < p < 0.39 + 0.025$

$0.365 < p < 0.415$

5.

$\hat{p} = \frac{X}{n} = \frac{55}{450} = 0.12$

$\hat{q} = 1 - 0.12 = 0.88$

$\hat{p} - (z_{\frac{\alpha}{2}})\sqrt{\frac{\hat{p}\hat{q}}{n}} < p < \hat{p} + (z_{\frac{\alpha}{2}})\sqrt{\frac{\hat{p}\hat{q}}{n}}$

$0.12 - 1.96\sqrt{\frac{(0.12)(0.88)}{450}} < p < 0.12 + 1.96\sqrt{\frac{(0.12)(0.88)}{450}}$

$0.12 - 0.03 < p < 0.12 + 0.03$

0.09 or $9\% < p < 0.15$ or 15%

(TI answer: $0.092 < p < 0.153$)

11% is contained in the confidence interval.

7.

$\hat{p} = 0.84$ $\hat{q} = 0.16$

$\hat{p} - (z_{\frac{\alpha}{2}})\sqrt{\frac{\hat{p}\hat{q}}{n}} < p < \hat{p} + (z_{\frac{\alpha}{2}})\sqrt{\frac{\hat{p}\hat{q}}{n}}$

7. continued

$0.84 - 1.65\sqrt{\frac{(0.84)(0.16)}{200}} < p <$
$\qquad 0.84 + 1.65\sqrt{\frac{(0.84)(0.16)}{200}}$

$0.84 - 0.043 < p < 0.84 + 0.043$

$0.797 < p < 0.883$

9.

$\hat{p} = 0.65$ $\hat{q} = 0.36$

$\hat{p} - (z_{\frac{\alpha}{2}})\sqrt{\frac{\hat{p}\hat{q}}{n}} < p < \hat{p} + (z_{\frac{\alpha}{2}})\sqrt{\frac{\hat{p}\hat{q}}{n}}$

$0.65 - 1.96\sqrt{\frac{(0.65)(0.35)}{300}} < p <$
$\qquad 0.65 + 1.96\sqrt{\frac{(0.65)(0.35)}{300}}$

$0.65 - 0.054 < p < 0.65 + 0.054$

$0.596 < p < 0.704$

11.

$\hat{p} = 0.25$ $\hat{q} = 0.75$

$\hat{p} - (z_{\frac{\alpha}{2}})\sqrt{\frac{\hat{p}\hat{q}}{n}} < p < \hat{p} + (z_{\frac{\alpha}{2}})\sqrt{\frac{\hat{p}\hat{q}}{n}}$

$0.25 - 2.58\sqrt{\frac{(0.25)(0.75)}{80}} < p <$
$\qquad 0.25 + 2.58\sqrt{\frac{(0.25)(0.75)}{80}}$

$0.25 - 0.125 < p < 0.25 + 0.125$

$0.125 < p < 0.375$ or $12.5\% < \mu < 37.5\%$

It would not be considered large since 0.28 is in the interval.

13.

$\hat{p} = 0.44975$ $\hat{q} = 0.55025$

$\hat{p} - (z_{\frac{\alpha}{2}})\sqrt{\frac{\hat{p}\hat{q}}{n}} < p < \hat{p} + (z_{\frac{\alpha}{2}})\sqrt{\frac{\hat{p}\hat{q}}{n}}$

$0.44975 - 1.96\sqrt{\frac{(0.44975)(0.55025)}{1005}} < p <$
$\qquad 0.44975 + 1.96\sqrt{\frac{(0.44975)(0.55025)}{1005}}$

$0.44975 - 0.03076 < p <$
$\qquad\qquad 0.44975 + 0.03076$

$0.419 < p < 0.481$

15.

$\hat{p} = \frac{40}{200} = 0.2$ $\hat{q} = 0.8$

$n = \hat{p}\,\hat{q}\left[\frac{z_{\frac{\alpha}{2}}}{E}\right]^2 = (0.2)(0.8)\left[\frac{1.96}{0.04}\right]^2$

$n = 384.16$ or 385

If no estimate of the sample proportion is available, use $\hat{p} = 0.5$:

$\hat{p} = 0.5$ $\hat{q} = 0.5$

$n = \hat{p}\,\hat{q}\left[\frac{z_{\frac{\alpha}{2}}}{E}\right]^2 = (0.5)(0.5)\left[\frac{1.96}{0.04}\right]^2$

$n = 600.25$ or 601

17.

a. $\hat{p} = 0.25 \qquad \hat{q} = 0.75$

$n = \hat{p}\,\hat{q}\left[\dfrac{z_{\frac{\alpha}{2}}}{E}\right]^2 = (0.25)(0.75)\left[\dfrac{1.96}{0.03}\right]^2$

$n = 800.33$ or 801

b. $\hat{p} = 0.5 \qquad\qquad \hat{q} = 0.5$

$n = \hat{p}\,\hat{q}\left[\dfrac{z_{\frac{\alpha}{2}}}{E}\right]^2 = (0.5)(0.5)\left[\dfrac{1.96}{0.03}\right]^2$

$n = 1067.11$ or 1068

19.

$\hat{p} = 0.5 \qquad\qquad \hat{q} = 0.5$

$n = \hat{p}\,\hat{q}\left[\dfrac{z_{\frac{\alpha}{2}}}{E}\right]^2$

$n = (0.5)(0.5)\left[\dfrac{1.65}{0.025}\right]^2$

$n = 1089$

21.

$600 = (0.5)(0.5)\left[\dfrac{z}{0.04}\right]^2$

$600 = 156.25 z^2$

$3.84 = z^2$

$\sqrt{3.84} = 1.96 = z$

1.96 corresponds to a 95% degree of confidence.

EXERCISE SET 7-4

1.

χ^2

3.

	χ^2_{left}	χ^2_{right}
a.	3.816	21.920
b.	10.117	30.144
c.	13.844	41.923
d.	0.412	16.750
e.	26.509	55.758

5.

$\dfrac{(n-1)s^2}{\chi^2_{\text{right}}} < \sigma^2 < \dfrac{(n-1)s^2}{\chi^2_{\text{left}}}$

$\dfrac{16(10.1)^2}{28.845} < \sigma^2 < \dfrac{16(10.1)^2}{6.908}$

$56.6 < \sigma^2 < 236.3$

$7.5 < \sigma < 15.4$

7.

Use the Range Rule of Thumb for σ.

$\sigma \approx \dfrac{\$25,462 - \$17,627}{4} = 1958.75$

7. continued

For insurer's cost:

$\dfrac{(n-1)s^2}{\chi^2_{\text{right}}} < \sigma^2 < \dfrac{(n-1)s^2}{\chi^2_{\text{left}}}$

$\dfrac{9(1958.75)^2}{21.666} < \sigma^2 < \dfrac{9(1958.75)^2}{2.088}$

$\$1,593,756 < \sigma^2 < \$16,537,507$

$\$1262.4 < \sigma < \$4066.6$

For uninsured person's cost:

$\sigma \approx \dfrac{\$58,702 - \$40,640}{4} = \$4515.5$

$\dfrac{9(4515.5)^2}{21.666} < \sigma^2 < \dfrac{9(4515.5)^2}{2.088}$

$\$8,469,845 < \sigma^2 < \$87,886,811$

$\$2910.3 < \sigma < \$9374.8$

9.

$s = 36.56$

$\dfrac{(n-1)s^2}{\chi^2_{\text{right}}} < \sigma^2 < \dfrac{(n-1)s^2}{\chi^2_{\text{left}}}$

$\dfrac{5(36.56)^2}{11.071} < \sigma^2 < \dfrac{5(36.56)^2}{1.145}$

$604 < \sigma^2 < 5837$

$24.6 < \sigma < 76.4$

11.

$\dfrac{(n-1)s^2}{\chi^2_{\text{right}}} < \sigma^2 < \dfrac{(n-1)s^2}{\chi^2_{\text{left}}}$

$\dfrac{24(462)^2}{39.364} < \sigma^2 < \dfrac{24(462)^2}{12.401}$

$130,135.6 < \sigma^2 < 413,084.1$

$360.8 < \sigma < 642.7$ or $361 < \sigma < 643$

13.

$s - z_{\frac{\alpha}{2}}\left(\dfrac{s}{\sqrt{2n}}\right) < \sigma < s + z_{\frac{\alpha}{2}}\left(\dfrac{s}{\sqrt{2n}}\right)$

$18 - 1.96\left(\dfrac{18}{\sqrt{400}}\right) < \sigma < 18 + 1.96\left(\dfrac{18}{\sqrt{400}}\right)$

$16.2 < \sigma < 19.8$

REVIEW EXERCISES - CHAPTER 7

1.

$\overline{X} = 19.63$ is the point estimate of μ.

$\overline{X} - z_{\frac{\alpha}{2}}\left(\dfrac{s}{\sqrt{n}}\right) < \mu < \overline{X} + z_{\frac{\alpha}{2}}\left(\dfrac{s}{\sqrt{n}}\right)$

$19.63 - 1.65\left(\dfrac{18.73}{\sqrt{30}}\right) < \mu <$

$\qquad\qquad 19.63 + 1.65\left(\dfrac{18.73}{\sqrt{30}}\right)$

1. continued
$19.63 - 5.64 < \mu < 19.63 + 5.64$
$13.99 < \mu < 25.27$ or $14 < \mu < 25$
(TI answer: $14.005 < \mu < 25.255$)

3.
$$n = \left[\frac{z_{\frac{\alpha}{2}} \sigma}{E} \right]^2 = \left[\frac{1.65(80)}{25} \right]^2$$

$$n = (5.28)^2 = 27.88 \text{ or } 28$$

5.
$\overline{X} = 82.64$ is the point estimate of μ.

$$\overline{X} - t_{\frac{\alpha}{2}} \left(\frac{s}{\sqrt{n}} \right) < \mu < \overline{X} + t_{\frac{\alpha}{2}} \left(\frac{s}{\sqrt{n}} \right)$$

$$82.64 - 2.228 \left(\frac{8.49}{\sqrt{11}} \right) < \mu <$$
$$82.64 + 2.228 \left(\frac{8.49}{\sqrt{11}} \right)$$

$82.64 - 5.7 < \mu < 82.64 + 5.7$
$76.9 < \mu < 88.3$

7.
$$\hat{p} - (z_{\frac{\alpha}{2}}) \sqrt{\frac{\hat{p}\hat{q}}{n}} < p < \hat{p} + (z_{\frac{\alpha}{2}}) \sqrt{\frac{\hat{p}\hat{q}}{n}}$$

$$0.44 - 1.96 \sqrt{\frac{(0.44)(0.56)}{1004}} < P <$$
$$0.44 + 1.96 \sqrt{\frac{(0.44)(0.56)}{1004}}$$

$0.44 - 0.0307 < p < 0.44 + 0.0307$

$0.409 < p < 0.471$

9.
$\hat{p} = \frac{80}{200} = 0.4 \qquad \hat{q} = 0.6$

$$\hat{p} - (z_{\frac{\alpha}{2}}) \sqrt{\frac{\hat{p}\hat{q}}{n}} < p < \hat{p} + (z_{\frac{\alpha}{2}}) \sqrt{\frac{\hat{p}\hat{q}}{n}}$$

$$0.4 - 1.65 \sqrt{\frac{(0.4)(0.6)}{200}} < p <$$
$$0.4 + 1.65 \sqrt{\frac{(0.4)(0.6)}{200}}$$

$0.4 - 0.057 < p < 0.4 + 0.057$
$0.343 < p < 0.457$

11.
$\hat{p} = 0.88 \qquad \hat{q} = 0.12$

$$n = \hat{p}\hat{q} \left[\frac{z_{\frac{\alpha}{2}}}{E} \right]^2 = (0.88)(0.12) \left[\frac{1.65}{0.025} \right]^2$$

$n = 459.99$ or 460

13.
$$\frac{(n-1)s^2}{\chi^2_{\text{right}}} < \sigma^2 < \frac{(n-1)s^2}{\chi^2_{\text{left}}}$$

$$\frac{(18-1)(0.29)^2}{30.191} < \sigma^2 < \frac{(18-1)(0.29)^2}{7.564}$$

$0.0474 < \sigma^2 < 0.1890$
$0.218 < \sigma < 0.435$
Yes; it seems that there is a large standard deviation.

15.
$$\frac{(n-1)s^2}{\chi^2_{\text{right}}} < \sigma^2 < \frac{(n-1)s^2}{\chi^2_{\text{left}}}$$

$$\frac{(15-1)(8.6)}{23.685} < \sigma^2 < \frac{(15-1)(8.6)}{6.571}$$

$5.1 < \sigma^2 < 18.3$

CHAPTER 7 QUIZ

1. True
2. True
3. False, it is consistent if, as sample size increases, the estimator approaches the parameter being estimated.
4. True
5. b
6. a
7. b
8. Unbiased, consistent, relatively efficient
9. Maximum error of estimate
10. Point
11. 90, 95, 99

12.
$\overline{X} = \$121.6$ is the point estimate for μ.
$$\overline{X} - z_{\frac{\alpha}{2}} \left(\frac{s}{\sqrt{n}} \right) < \mu < \overline{X} + z_{\frac{\alpha}{2}} \left(\frac{s}{\sqrt{n}} \right)$$

$$\$121.60 - 1.65 \left(\frac{6.36}{\sqrt{36}} \right) < \mu <$$
$$\$121.60 + 1.65 \left(\frac{6.36}{\sqrt{36}} \right)$$

$\$119.85 < \mu < \$123.35$

13.
$\overline{X} = \$44.80$ is the point estimate for μ.

$$\overline{X} - t_{\frac{\alpha}{2}} \left(\frac{s}{\sqrt{n}} \right) < \mu < \overline{X} + t_{\frac{\alpha}{2}} \left(\frac{s}{\sqrt{n}} \right)$$

$$\$44.80 - 2.093 \left(\frac{3.53}{\sqrt{20}} \right) < \mu <$$
$$\$44.80 + 2.093 \left(\frac{3.53}{\sqrt{20}} \right)$$

$\$43.15 < \mu < \$46.45$

14.

$\overline{X} = 4150$ is the point estimate for μ.

$$\overline{X} - z_{\frac{\alpha}{2}}\left(\frac{s}{\sqrt{n}}\right) < \mu < \overline{X} + z_{\frac{\alpha}{2}}\left(\frac{s}{\sqrt{n}}\right)$$

$$\$4150 - 2.58\left(\frac{480}{\sqrt{40}}\right) < \mu <$$
$$\$4150 + 2.58\left(\frac{480}{\sqrt{40}}\right)$$

$$\$3954 < \mu < \$4346$$

15.

$$\overline{X} - t_{\frac{\alpha}{2}}\left(\frac{s}{\sqrt{n}}\right) < \mu < \overline{X} + t_{\frac{\alpha}{2}}\left(\frac{s}{\sqrt{n}}\right)$$

$$48.6 - 2.262\left(\frac{4.1}{\sqrt{10}}\right) < \mu < 48.6 + 2.262\left(\frac{4.1}{\sqrt{10}}\right)$$

$$45.7 < \mu < 51.5$$

16.

$$\overline{X} - t_{\frac{\alpha}{2}}\left(\frac{s}{\sqrt{n}}\right) < \mu < \overline{X} + t_{\frac{\alpha}{2}}\left(\frac{s}{\sqrt{n}}\right)$$

$$438 - 3.499\left(\frac{16}{\sqrt{8}}\right) < \mu < 438 + 3.499\left(\frac{16}{\sqrt{8}}\right)$$

$$418 < \mu < 458$$

17.

$$\overline{X} - t_{\frac{\alpha}{2}}\left(\frac{s}{\sqrt{n}}\right) < \mu < \overline{X} + t_{\frac{\alpha}{2}}\left(\frac{s}{\sqrt{n}}\right)$$

$$31 - 2.353\left(\frac{4}{\sqrt{4}}\right) < \mu < 31 + 2.353\left(\frac{4}{\sqrt{4}}\right)$$

$$26 < \mu < 36$$

18.

$$n = \left[\frac{z_{\frac{\alpha}{2}}\,\sigma}{E}\right]^2 = \left[\frac{2.58(2.6)}{0.5}\right]^2$$
$$n = 179.98 \text{ or } 180$$

19.

$$n = \left[\frac{z_{\frac{\alpha}{2}}\,\sigma}{E}\right]^2 = \left[\frac{1.65(900)}{300}\right]^2$$
$$n = 24.5 \text{ or } 25$$

20.

$$\hat{p} - (z_{\frac{\alpha}{2}})\sqrt{\frac{\hat{p}\,\hat{q}}{n}} < p < \hat{p} + (z_{\frac{\alpha}{2}})\sqrt{\frac{\hat{p}\,\hat{q}}{n}}$$

$$\hat{p} = 0.43 \qquad \hat{q} = 0.57$$

$$0.43 - 1.96\sqrt{\frac{(0.43)(0.57)}{300}} < p <$$
$$0.43 + 1.96\sqrt{\frac{(0.43)(0.57)}{300}}$$

$$0.374 < p < 0.486$$

21.

$$\hat{p} - (z_{\frac{\alpha}{2}})\sqrt{\frac{\hat{p}\,\hat{q}}{n}} < p < \hat{p} + (z_{\frac{\alpha}{2}})\sqrt{\frac{\hat{p}\,\hat{q}}{n}}$$

$$0.36 - 1.65\sqrt{\frac{(0.36)(0.64)}{150}} < p <$$
$$0.36 + 1.65\sqrt{\frac{(0.36)(0.64)}{150}}$$

$$0.295 < p < 0.425$$

22.

$$\hat{p} - (z_{\frac{\alpha}{2}})\sqrt{\frac{\hat{p}\,\hat{q}}{n}} < p < \hat{p} + (z_{\frac{\alpha}{2}})\sqrt{\frac{\hat{p}\,\hat{q}}{n}}$$

$$0.4444 - 1.96\sqrt{\frac{(0.4444)(0.5556)}{90}} < p <$$
$$0.4444 + 1.96\sqrt{\frac{(0.4444)(0.5556)}{90}}$$

$$0.342 < p < 0.547$$

23.

$$n = \hat{p}\,\hat{q}\left[\frac{z_{\frac{\alpha}{2}}}{E}\right]^2$$

$$n = (0.15)(0.85)\left[\frac{1.96}{0.03}\right]^2$$

$$n = 544.22 \text{ or } 545$$

24.

$$\frac{(n-1)s^2}{\chi^2_{\text{right}}} < \sigma^2 < \frac{(n-1)s^2}{\chi^2_{\text{left}}}$$

$$\frac{24(9)^2}{39.364} < \sigma^2 < \frac{24(9)^2}{12.401}$$

$$49.4 < \sigma^2 < 156.8$$
$$7 < \sigma < 13$$

25.

$$\frac{(n-1)s^2}{\chi^2_{\text{right}}} < \sigma^2 < \frac{(n-1)s^2}{\chi^2_{\text{left}}}$$

$$\frac{26(6.8)^2}{38.885} < \sigma^2 < \frac{26(6.8)^2}{15.379}$$

$$30.9 < \sigma^2 < 78.2$$
$$5.6 < \sigma < 8.8$$

26.

$$\frac{(n-1)s^2}{\chi^2_{\text{right}}} < \sigma^2 < \frac{(n-1)s^2}{\chi^2_{\text{left}}}$$

$$\frac{19(2.3)^2}{30.144} < \sigma^2 < \frac{19(2.3)^2}{10.177}$$

$$3.33 < \sigma^2 < 10$$
$$1.8 < \sigma < 3.2$$

Note: Graphs are not to scale and are intended to convey a general idea. Answers may vary due to rounding.

EXERCISE SET 8-1

1.
The null hypothesis is a statistical hypothesis that states there is no difference between a parameter and a specific value or there is no difference between two parameters. The alternative hypothesis specifies a specific difference between a parameter and a specific value, or that there is a difference between two parameters. Examples will vary.

3.
A statistical test uses the data obtained from a sample to make a decision as to whether or not the null hypothesis should be rejected.

5.
The critical region is the region of values of the test-statistic that indicates a significant difference and the null hypothesis should be rejected. The non-critical region is the region of values of the test-statistic that indicates the difference was probably due to chance, and the null hypothesis should not be rejected.

7.
Type I is represented by α, type II is represented by β.

9.
A one-tailed test should be used when a specific direction, such as greater than or less than, is being hypothesized, whereas when no direction is specified, a two-tailed test should be used.

11.
Hypotheses can only be proved true when the entire population is used to compute the test statistic. In most cases, this is impossible.

12.
a. ± 1.96

b. -2.33

c. $+2.58$

d. $+2.33$

e. -1.65

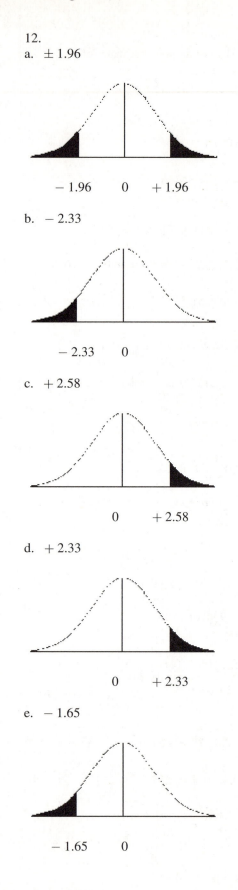

$-1.96 \quad 0 \quad +1.96$

$-2.33 \quad 0$

$0 \quad +2.58$

$0 \quad +2.33$

$-1.65 \quad 0$

f. -2.05

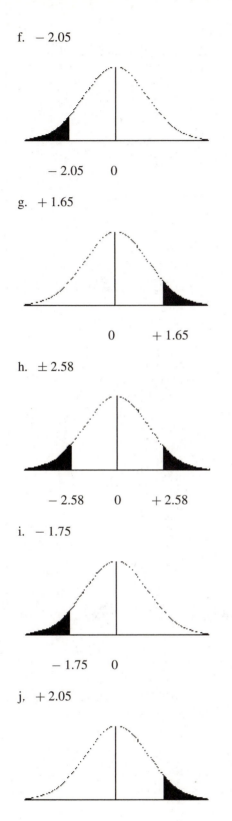

-2.05 0

g. $+1.65$

0    $+1.65$

h. ± 2.58

-2.58 0 $+2.58$

i. -1.75

-1.75 0

j. $+2.05$

0    $+2.05$

13.

a. H_0: $\mu = 24.6$
    H_1: $\mu \neq 24.6$

b. H_0: $\mu = \$51,497$
    H_1: $\mu \neq \$51,497$

c. H_0: $\mu = 25.4$
    H_1: $\mu > 25.4$

d. H_0: $\mu = 88$
    H_1: $\mu < 88$

e. H_0: $\mu = 70$
    H_1: $\mu < 70$

f. H_0: $\mu = \$79.95$
    H_1: $\mu \neq \$79.95$

g. H_0: $\mu = 8.2$
    H_1: $\mu \neq 8.2$

EXERCISE SET 8-2

1.
H_0: $\mu = 305$
H_1: $\mu > 305$ (claim)

C. V. $= 1.645$ $\sigma = \sqrt{3.57} = 1.89$
$z = \dfrac{\overline{X} - \mu}{\frac{\sigma}{\sqrt{n}}} = \dfrac{306.2 - 305}{\frac{1.89}{\sqrt{55}}} = 4.71$

0    $+1.645$ $\uparrow$
                 4.71

Reject the null hypothesis. There is enough evidence to support the claim that the average depth has increased. Many factors could contribute to the increase, including warmer temperatures and higher than usual rainfall.

3.
H_0: $\mu = \$24$ billion
$H1$: $\mu > \$24$ billion   (claim)

C. V. $= +1.65$ $\overline{X} = \$31.5$ $s = \$28.7$
$z = \dfrac{\overline{X} - \mu}{\frac{\sigma}{\sqrt{n}}} = \dfrac{31.5 - 24}{\frac{28.7}{\sqrt{50}}} = 1.85$

3. continued

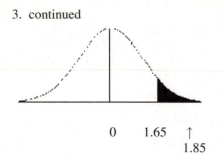

0   1.65   ↑
              1.85

Reject the null hypothesis. There is enough evidence to support the claim that the average revenue exceeds $24 billion.

5.
H_0: $\mu = 30.9$
H_1: $\mu \neq 30.9$ (claim)

C. V. $= \pm 2.58$
$z = \frac{\overline{X} - \mu}{\frac{s}{\sqrt{n}}} = \frac{32.1 - 30.9}{\frac{3.6}{\sqrt{32}}} = 1.89$

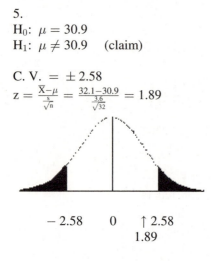

-2.58 0   ↑ 2.58
                  1.89

Do not reject the null hypothesis. There is not enough evidence to support the claim that average number of hours differs from 30.9.

7.
H_0: $\mu = 29$
H_1: $\mu \neq 29$ (claim)

C. V. $= \pm 1.96$ $\overline{X} = 29.45$ $s = 2.61$
$z = \frac{\overline{X} - \mu}{\frac{\sigma}{\sqrt{n}}} = \frac{29.45 - 29}{\frac{2.61}{\sqrt{30}}} = 0.944$

1.96   0   ↑ 1.96
                0.944

7. continued
Do not reject the null hypothesis. There is enough evidence to reject the claim that the average height differs from 29 inches.

9.
H_0: $\mu = \$8121$
H_1: $\mu > \$8121$ (claim)

C. V. $= 2.326$
$z = \frac{\overline{X} - \mu}{\frac{\sigma}{\sqrt{n}}} = \frac{\$8350 - \$8121}{\frac{750}{\sqrt{40}}} = 1.93$

0   ↑ 2.326
1.93

Do not reject the null hypothesis. There is not enough evidence to support the claim that the average operating cost has increased.

11.
H_0: $\mu = 500$
H_1: $\mu \neq 500$ (claim)

C. V. $= \pm 2.58$
$z = \frac{\overline{X} - \mu}{\frac{\sigma}{\sqrt{n}}} = \frac{476 - 500}{\frac{42}{\sqrt{50}}} = -4.04$

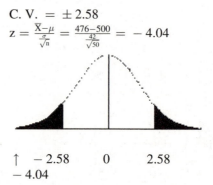

↑ -2.58 0   2.58
-4.04

Reject the null hypothesis. There is enough evidence to conclude that the average number of cups differs from 500.

13.
H_0: $\mu = 60.35$
H_1: $\mu < 60.35$ (claim)

C. V. $= -1.65$
$z = \frac{\overline{X} - \mu}{\frac{s}{\sqrt{n}}} = \frac{55.4 - 60.35}{\frac{6.5}{\sqrt{40}}} = -4.82$

13. continued

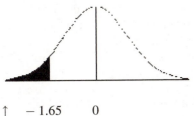

↑ -1.65 0
-4.82

Reject the null hypothesis. There is enough evidence to support the claim that state senators are younger than U. S. Senators.

15.
a. Do not reject.
b. Reject.
c. Do not reject.
d. Reject.
e. Reject.

17.
H_0: $\mu = 264$
H_1: $\mu < 264$ (claim)
$z = \frac{\overline{X} - \mu}{\frac{\sigma}{\sqrt{n}}} = \frac{262.3 - 264}{\frac{3}{\sqrt{20}}} = -2.53$

The area corresponding to z = 2.53 is 0.9943. The P-value is $1 - 0.9943 = 0.0057$. The decision is to reject the null hypothesis since $0.0057 < 0.01$. There is enough evidence to support the claim that the average stopping distance is less than 264 feet.
(TI: P-value = 0.0056)

19.
H_0: $\mu = 546$
H_1: $\mu < 546$ (claim)
$z = \frac{\overline{X} - \mu}{\frac{\sigma}{\sqrt{n}}} = \frac{544.8 - 546}{\frac{3}{\sqrt{36}}} = -2.4$

The area corresponding to z = -2.4 is 0.0082. Thus, P-value = 0.0082. The decision is to reject the null hypothesis since $0.0082 < 0.01$. There is enough evidence to support the claim that the number of calories burned is less than 546.
(TI: P-value = 0.0082)

21.
H_0: $\mu = 444$
H_1: $\mu \neq 444$
$z = \frac{\overline{X} - \mu}{\frac{\sigma}{\sqrt{n}}} = \frac{430 - 444}{\frac{52}{\sqrt{40}}} = -1.70$

21. continued
The area corresponding to z = -1.70 is 0.0446. To get the P-value, multiply by 2 since this is a two-tailed test. Thus, P-value = $2(0.0446) = 0.0892$. The decision is do not reject the null hypothesis since P-value > 0.05. There is not enough evidence to support the claim that the mean differs from 444.
(TI: P-value = 0.0886)

23.
H_0: $\mu = 30,000$ (claim)
H_1: $\mu \neq 30,000$
$z = \frac{\overline{X} - \mu}{\frac{s}{\sqrt{n}}} = \frac{30,456 - 30,000}{\frac{1684}{\sqrt{40}}} = 1.71$

The area corresponding to z = 1.71 is 0.9564. The P-value is $2(1 - 0.9564) = 2(0.0436) = 0.0872$. The decision is to reject the null hypothesis at $\alpha = 0.10$ since $0.0872 < 0.10$. The conclusion is that there is enough evidence to reject the claim that customers are adhering to the recommendation. A 0.10 significance level is probably appropriate since there is little consequence of a Type I error. The dealer would be advised to increase efforts to make its customers aware of the service recommendation.
(TI: P-value = 0.0868)

25.
H_0: $\mu = 10$
H_1: $\mu < 10$ (claim)

$\overline{X} = 5.025$ s = 3.63
$z = \frac{\overline{X} - \mu}{\frac{s}{\sqrt{n}}} = \frac{5.025 - 10}{\frac{3.63}{\sqrt{40}}} = -8.67$

The area corresponding to 8.67 is greater than 0.9999. The P-value is $1 - 0.9999 < 0.0001$. Since $0.0001 < 0.05$, the decision is to reject the null hypothesis. There is enough evidence to support the claim that the average number of days missed per year is less than 10.

27.
The mean and standard deviation are found as follows:

27. continued

	f	X_m	$f \cdot X_m$	$f \cdot X_m^2$
8.35 - 8.43	2	8.39	16.78	140.7842
8.44 - 8.52	6	8.48	50.88	431.4624
8.53 - 8.61	12	8.57	102.84	881.3388
8.62 - 8.70	18	8.66	155.88	1349.9208
8.71 - 8.79	10	8.75	87.5	765.625
8.80 - 8.88	2	8.84	17.68	156.2912
	50		431.56	3725.4224

$$\overline{X} = \frac{\sum f \cdot X_m}{n} = \frac{431.56}{50} = 8.63$$

$$s = \sqrt{\frac{\sum f \cdot X_m^2 - \frac{\sum(f \cdot X_m)^2}{n}}{n-1}} = \sqrt{\frac{3725.4224 - \frac{(431.56)^2}{50}}{49}}$$

$$= 0.105$$

H_0: $\mu = 8.65$ (claim)
H_1: $\mu \neq 8.65$

C. V. $= \pm 1.96$
$$z = \frac{\overline{X} - \mu}{\frac{s}{\sqrt{n}}} = \frac{8.63 - 8.65}{\frac{0.105}{\sqrt{50}}} = -1.35$$

Do not reject the null hypothesis. There is not enough evidence to reject the claim that the average hourly wage of the employees is $8.65.

EXERCISE SET 8-3

1.
It is bell-shaped, symmetric about the mean, and it never touches the x axis. The mean, median, and mode are all equal to 0 and they are located at the center of the distribution. The t distribution differs from the standard normal distribution in that it is a family of curves, the variance is greater than one, and as the degrees of freedom increase the t distribution approaches the standard normal distribution.

3.
a. d. f. $= 9$ C. V. $= +1.833$
b. d. f. $= 17$ C. V. $= \pm 1.740$
c. d. f. $= 5$ C. V. $= -3.365$
d. d. f. $= 8$ C. V. $= +2.306$
e. d. f. $= 14$ C. V. $= \pm 2.145$
f. d. f. $= 22$ C. V. $= -2.819$
g. d. f. $= 27$ C. V. $= \pm 2.771$
h. d. f. $= 16$ C. V. $= \pm 2.583$

4.
a. $0.01 <$ P-value < 0.025 (0.018)

4. continued
b. $0.05 <$ P-value < 0.10 (0.062)
c. $0.10 <$ P-value < 0.25 (0.123)
d. $0.10 <$ P-value < 0.20 (0.138)
e. P-value < 0.005 (0.003)
f. $0.10 <$ P-value < 0.25 (0.158)
g. P-value $= 0.05$ (0.05)
h. P-value > 0.25 (0.261)

5.
H_0: $\mu = 179$
H_1: $\mu \neq 179$ (claim)

C. V. $= \pm 3.250$ d. f. $= 9$

$$t = \frac{\overline{X} - \mu}{\frac{s}{\sqrt{n}}} = \frac{205 - 179}{\frac{26}{\sqrt{10}}} = 3.162$$

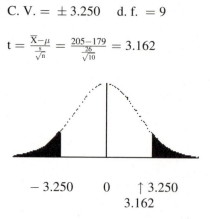

-3.250 0 $\uparrow 3.250$
3.162

Do not reject the null hypothesis. There is not enough evidence to support the claim that the average expense differs from $179.

7.
H_0: $\mu = 2.27$
H_1: $\mu \neq 2.27$ (claim)

C. V. $= \pm 2.093$ d. f. $= 19$
$$t = \frac{\overline{X} - \mu}{\frac{s}{\sqrt{n}}} = \frac{2.98 - 2.27}{\frac{0.98}{\sqrt{20}}} = 3.24$$

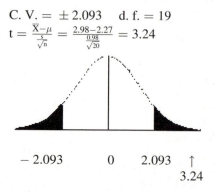

-2.093 0 2.093 $\uparrow$
3.24

Reject the null hypothesis. There is enough evidence to support the claim that the average call differs from 2.27 minutes.

9.
H_0: $\mu = 700$ (claim)
H_1: $\mu < 700$

9. continued
$\overline{X} = 606.5 \quad s = 109.1$

C. V. $= -2.262 \quad$ d. f. $= 9$
$t = \frac{\overline{X} - \mu}{\frac{s}{\sqrt{n}}} = \frac{606.5 - 700}{\frac{109.1}{\sqrt{10}}} = -2.71$

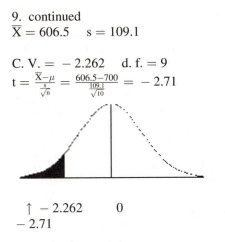

$\uparrow -2.262 \qquad 0$
-2.71

Reject the null hypothesis. There is enough evidence to reject the claim that the average height of the buildings is at least 700 feet.

11.
$H_0: \mu = 73$
$H_1: \mu > 73 \quad$ (claim)

C. V. $= 2.821 \quad$ d. f. $= 9$
$\overline{X} = 123.5 \qquad s = 39.303$
$t = \frac{\overline{X} - \mu}{\frac{s}{\sqrt{n}}} = \frac{123.5 - 73}{\frac{39.303}{\sqrt{10}}} = 4.063$

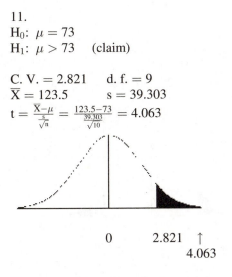

$0 \qquad 2.821 \quad \uparrow$
4.063

Reject the null hypothesis. There is enough evidence to support the claim that the average viewing time is longer than 73 minutes.

13.
$H_0: \mu = \$54.8$
$H_1: \mu > \$54.8 \quad$ (claim)

C. V. $= 1.761 \qquad$ d. f. $= 14$
$t = \frac{\overline{X} - \mu}{\frac{s}{\sqrt{n}}} = \frac{\$62.3 - \$54.8}{\frac{\$9.5}{\sqrt{15}}} = 3.058$

13. continued

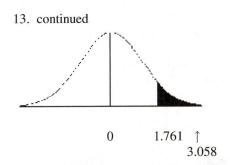

$0 \qquad 1.761 \quad \uparrow$
3.058

Reject the null hypothesis. There is enough evidence to support the claim that the cost to produce an action movie is more than $54.8 million.

15.
$H_0: \mu = \$50.07$
$H_1: \mu > \$50.07 \quad$ (claim)

C. V. $= 1.833 \quad$ d. f. $= 9$
$\overline{X} = \$56.11 \qquad s = \$6.97$
$t = \frac{\overline{X} - \mu}{\frac{s}{\sqrt{n}}} = \frac{\$56.11 - \$50.07}{\frac{6.97}{\sqrt{10}}} = 2.741$

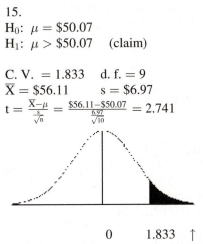

$0 \qquad 1.833 \quad \uparrow$
2.741

Reject the null hypothesis. There is enough evidence to support the claim that the average bill has increased.

17.
$H_0: \mu = 5.8$
$H_1: \mu \neq 5.8 \quad$ (claim)
$\overline{X} = 3.85 \quad s = 2.519$
d. f. $= 19 \quad \alpha = 0.05$
P-value $< 0.01 \quad (0.0026)$
$t = \frac{\overline{X} - \mu}{\frac{s}{\sqrt{n}}} = \frac{3.85 - 5.8}{\frac{2.519}{\sqrt{20}}} = -3.462$

Since P-value < 0.01, reject the null hypothesis. There is enough evidence to support the claim that the mean is not 5.8.

19.
$H_0: \mu = \$15,000$
$H_1: \mu \neq \$15,000 \quad$ (claim)

$\overline{X} = \$14,347.17 \qquad s = \$2048.54$

79

19. continued

d. f. $= 11$ C. V. $= \pm 2.201$

$$t = \frac{\overline{X}-\mu}{\frac{s}{\sqrt{n}}} = \frac{\$14,347.17 - \$15,000}{\frac{\$2048.54}{\sqrt{12}}} = -1.10$$

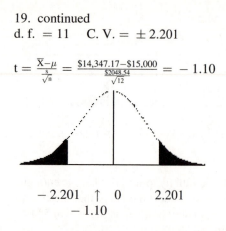

$$-2.201 \quad \uparrow \quad 0 \qquad 2.201$$
$$-1.10$$

Do not reject the null hypothesis. There is not enough evidence to say that the average stipend differs from $15,000.

EXERCISE SET 8-4

1.
Answers will vary.

3.
$np \geq 5$ and $nq \geq 5$

5.
H_0: $p = 0.686$
H_1: $p \neq 0.686$ (claim)

$\hat{p} = \frac{92}{150} = 0.613$ $p = 0.686$ $q = 0.314$
C. V. $= \pm 2.58$

$$z = \frac{\hat{p}-p}{\sqrt{\frac{pq}{n}}} = \frac{0.613-0.686}{\sqrt{\frac{(0.686)(0.314)}{150}}} = -1.93$$
(TI: $z = -1.918$)

$$-2.58 \quad \uparrow \quad 0 \qquad 2.58$$
$$-1.93$$

Do not reject the null hypothesis. There is not enough evidence to support the claim that the proportion of homeowners is different from 68.6%.

7.
H_0: $p = 0.188$
H_1: $p < 0.188$ (claim)

7. continued

$\hat{p} = \frac{20}{130} = 0.1538$ $p = 0.188$ $q = 0.812$
C. V. $= -1.645$
$$z = \frac{\hat{p}-p}{\sqrt{\frac{pq}{n}}} = \frac{0.1538-0.188}{\sqrt{\frac{(0.188)(0.812)}{130}}} = -0.998 \text{ or } -1.0$$
(TI answer $= -0.997$)

$$-1.645 \quad \uparrow \quad 0$$
$$-1.00$$

Do not reject the null hypothesis. There is not enough evidence to conclude that the proportion is less than 18.8%.

9.
H_0: $p = 0.47$
H_1: $p \neq 0.47$ (claim)

$\hat{p} = \frac{263}{500} = 0.526$ $p = 0.47$ $q = 0.53$
C. V. $= \pm 1.96$
$$z = \frac{\hat{p}-p}{\sqrt{\frac{pq}{n}}} = \frac{0.526-0.47}{\sqrt{\frac{(0.47)(0.53)}{500}}} = 2.509 \text{ or } 2.51$$

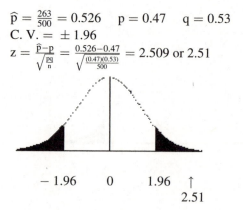

$$-1.96 \qquad 0 \qquad 1.96 \quad \uparrow$$
$$2.51$$

Reject the null hypothesis. There is enough evidence to support the claim that the percentage of women is not 47%.

11.
H_0: $p = 0.32$
H_1: $p \neq 0.32$ (claim)

$\hat{p} = \frac{170}{423} = 0.4019$ $p = 0.32$ $q = 0.68$
C. V. $= \pm 2.576$
$$z = \frac{\hat{p}-p}{\sqrt{\frac{pq}{n}}} = \frac{0.4019-0.32}{\sqrt{\frac{(0.32)(0.68)}{423}}} = 3.611$$

11. continued

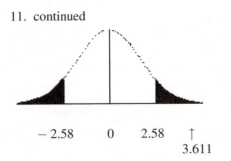

$$-2.58 \quad 0 \quad 2.58 \quad \uparrow$$
$$3.611$$

Reject the null hypothesis. There is enough evidence to support the claim that the proportion of 40- to 59-year-old moviegoers is not 32%.

13.
H_0: $p = 0.54$ (claim)
H_1: $p \neq 0.54$

$\hat{p} = \frac{36}{60} = 0.6 \quad p = 0.54 \quad q = 0.46$
$z = \frac{\hat{p} - p}{\sqrt{\frac{pq}{n}}} = \frac{0.6 - 0.54}{\sqrt{\frac{(0.54)(0.46)}{60}}} = 0.93$
Area = 0.8238
P-value = $2(1 - 0.8238) = 0.3524$
Since P-value > 0.01, do not reject the null hypothesis. There is enough evidence to support the claim that 54% of kids had a snack after school. Yes, a healthy snack should be made available for children to eat after school.
(TI: P-value = 0.3511)

15.
H_0: $p = 0.18$ (claim)
H_1: $p \neq 0.18$

$\hat{p} = \frac{50}{300} = 0.1667 \quad p = 0.18 \quad q = 0.82$
$z = \frac{\hat{p} - p}{\sqrt{\frac{pq}{n}}} = \frac{0.1667 - 0.18}{\sqrt{\frac{(0.18)(0.82)}{300}}} = -0.60$
Area = 0.2743
P-value = $2(0.2743) = 0.5486$
Since P-value > 0.05, do not reject the null hypothesis. There is not enough evidence to reject the claim that 18% of all high school students smoke at least a pack of cigarettes a day. (TI: P-value = 0.5478)

17.
H_0: $p = 0.67$
H_1: $p \neq 0.67$ (claim)

$\hat{p} = \frac{82}{100} = 0.82 \quad p = 0.67 \quad q = 0.33$
C. V. = ± 1.96

17. continued
$z = \frac{\hat{p} - p}{\sqrt{\frac{pq}{n}}} = \frac{0.82 - 0.67}{\sqrt{\frac{(0.67)(0.33)}{100}}} = 3.19$

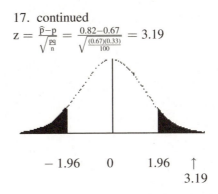

$$-1.96 \quad 0 \quad 1.96 \quad \uparrow$$
$$3.19$$

Reject the null hypothesis. There is enough evidence to support the claim that the percentage is not 67%.

19.
H_0: $p = 0.576$
H_1: $p < 0.576$ (claim)

$\hat{p} = \frac{17}{36} = 0.472 \quad p = 0.576 \quad q = 0.424$
C. V. = -1.65
$z = \frac{\hat{p} - p}{\sqrt{\frac{pq}{n}}} = \frac{0.472 - 0.576}{\sqrt{\frac{(0.576)(0.424)}{36}}} = -1.26$

$$-1.65 \quad \uparrow \quad 0$$
$$-1.26$$

Do not reject the null hypothesis. There is not enough evidence to support the claim that the percentage of injuries during practice is less than 57.6%.

21.
This represents a binomial distribution with $p = 0.50$ and $n = 9$. The P-value is $2 \cdot P(X \leq 3) = 2(0.254) = 0.508$.

Since P-value > 0.10, the conclusion that the coin is not balanced is probably false. The answer is no.

23.
$z = \frac{X - \mu}{\sigma}$

$z = \frac{X - np}{\sqrt{npq}}$

$z = \frac{\frac{X}{n} - \frac{np}{n}}{\frac{1}{n}\sqrt{npq}}$

23. continued

$$z = \frac{\frac{X}{n} - \frac{np}{n}}{\sqrt{\frac{npq}{n^2}}}$$

$$z = \frac{\hat{p} - p}{\sqrt{\frac{pq}{n}}}$$

EXERCISE SET 8-5

1.
a. H_0: $\sigma^2 = 225$
   H_1: $\sigma^2 > 225$

C. V. = 27.587   d. f. = 17

0   27.587

b. H_0: $\sigma^2 = 225$
   H_1: $\sigma^2 < 225$

C. V. = 14.042   d. f. = 22

0   14.042

c. H_0: $\sigma^2 = 225$
   H_1: $\sigma^2 \neq 225$

C. V. = 5.629, 26.119   d. f. = 14

0   5.629   26.119

d. H_0: $\sigma^2 = 225$
   H_1: $\sigma^2 \neq 225$

C. V. = 2.167, 14.067   d. f. = 7

1d. continued

0   2.167   14.067

e. H_0: $\sigma^2 = 225$
   H_1: $\sigma^2 > 225$

C. V. = 32.000   d. f. = 16

0   32.000

f. H_0: $\sigma^2 = 225$
   H_1: $\sigma^2 < 225$

C. V. = 8.907   d. f. = 19

0   8.907

g. H_0: $\sigma^2 = 225$
   H_1: $\sigma^2 \neq 225$

C. V. = 3.074, 28.299   d. f. = 12

0   3.074   28.299

h. H_0: $\sigma^2 = 225$
   H_1: $\sigma^2 < 225$

C. V. = 15.308   d. f. = 28

1h. continued

 0   15.308

2.
a. $0.01 < \text{P-value} < 0.025$ (0.015)
b. $0.005 < \text{P-value} < 0.01$ (0.006)
c. $0.01 < \text{P-value} < 0.025$ (0.012)
d. $\text{P-value} < 0.005$ (0.003)
e. $0.025 < \text{P-value} < 0.05$ (0.037)
f. $0.05 < \text{P-value} < 0.10$ (0.088)
g. $0.05 < \text{P-value} < 0.10$ (0.066)
h. $\text{P-value} < 0.01$ (0.007)

3.
H_0: $\sigma = 60$ (claim)
H_1: $\sigma \neq 60$

C. V. = 8.672, 27.587 $\alpha = 0.10$
d. f. = 17
s = 64.6
$\chi^2 = \frac{(n-1)s^2}{\sigma^2} = \frac{(18-1)(64.6)^2}{(60)^2} = 19.707$

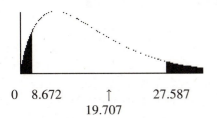

 0   8.672   ↑   27.587
                  19.707

Do not reject the null hypothesis. There is not enough evidence to reject the claim that the standard deviation is 60.

5.
H_0: $\sigma = 15$
H_1: $\sigma < 15$ (claim)

C. V. = 4.575 $\alpha = 0.05$ d. f. = 11

$\chi^2 = \frac{(n-1)s^2}{\sigma^2} = \frac{(12-1)(13.6)^2}{15^2} = 9.0425$

5. continued

 0 4.575   ↑
              9.0425

Do not reject the null hypothesis. There is not enough evidence to support the claim that the standard deviation is less than 15.

7.
H_0: $\sigma = 1.2$ (claim)
H_1: $\sigma > 1.2$

$\alpha = 0.01$ d. f. = 14

$\chi^2 = \frac{(n-1)s^2}{\sigma^2} = \frac{(15-1)(1.8)^2}{(1.2)^2} = 31.5$

P-value < 0.005   (0.0047)
Since P-value < 0.01, reject the null hypothesis. There is enough evidence to reject the claim that the standard deviation is less than or equal to 1.2 minutes.

9.
H_0: $\sigma = 100$
H_1: $\sigma > 100$ (claim)

s = 126.721
C. V. = 12.017 $\alpha = 0.10$ d. f. = 7

$\chi^2 = \frac{(n-1)s^2}{\sigma^2} = \frac{(8-1)(126.721)^2}{100^2} = 11.241$

 0   ↑ 12.017
                              11.241

Do not reject the null hypothesis. There is not enough evidence to support the claim that the standard deviation is more than 100 mg.

11.
H_0: $\sigma = 35$
H_1: $\sigma < 35$ (claim)

11. continued
C. V. $= 3.940$ $\alpha = 0.05$ d. f. $= 10$

$$\chi^2 = \frac{(n-1)s^2}{\sigma^2} = \frac{(11-1)(32)^2}{35^2} = 8.359$$

0   3.940   ↑
              8.359

Do not reject the null hypothesis. There is not enough evidence to support the claim that the standard deviation is less than 35.

13.
$\sigma \approx \frac{Range}{4}$
$\sigma \approx \frac{\$9500 - \$6782}{4} = \$679.50$
H_0: $\sigma = \$679.50$
H_1: $\sigma \neq \$679.50$ (claim)

$s = 770.67$
C. V. $= 5.009, 24.736$ $\alpha = 0.05$ d. f. $= 13$

$$\chi^2 = \frac{(n-1)s^2}{\sigma^2} = \frac{(14-1)(770.67)^2}{679.5^2} = 16.723$$

0   5.009   ↑   24.736
                   16.723

Do not reject the null hypothesis. There is not enough evidence to support the claim that the standard deviation is not $679.50.

15.
H_0: $\sigma = 0.52$
H_1: $\sigma > 0.52$ (claim)

C. V. $= 30.144$ $\alpha = 0.05$ d. f. $= 19$

$$\chi^2 = \frac{(n-1)s^2}{\sigma^2} = \frac{(20-1)(0.568)^2}{(0.52)^2} = 22.670$$

15. continued

0   ↑   30.144
                  22.670

Do not reject the null hypothesis. There is not enough evidence to support the claim that the standard deviation is more than 0.52 mm.

EXERCISE SET 8-6

1.
H_0: $\mu = \$273$
H_1: $\mu \neq \$273$ (claim)

C. V. $= \pm 1.96$
$z = \frac{\overline{X} - \mu}{\frac{\sigma}{\sqrt{n}}} = \frac{\$285 - \$273}{\frac{58}{\sqrt{40}}} = 1.31$

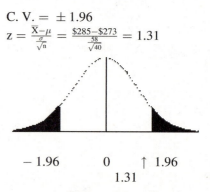

-1.96 0   ↑ 1.96
                      1.31

Do not reject the null hypothesis. There is not enough evidence to support the claim that the mean differs from $273.

The 95% confidence interval of the mean is:
$\overline{X} - z_{\frac{\alpha}{2}} \frac{\sigma}{\sqrt{n}} < \mu < \overline{X} + z_{\frac{\alpha}{2}} \frac{\sigma}{\sqrt{n}}$

$285 - 1.96 \left(\frac{58}{\sqrt{40}}\right) < \mu <$
$\qquad\qquad\qquad 285 + 1.96 \left(\frac{58}{\sqrt{40}}\right)$
$\$267.03 < \mu < \$302.97$

The hypothesized mean is within the interval, thus we can be 95% confident that the mean will be between $267.03 and $302.97.

3.
H_0: $\mu = \$19,150$
H_1: $\mu \neq \$19,150$ (claim)

3. continued
C. V. = ± 1.96
$$z = \frac{\overline{X} - \mu}{\frac{\sigma}{\sqrt{n}}} = \frac{\$17,020 - \$19,150}{\frac{4080}{\sqrt{50}}} = -3.692$$

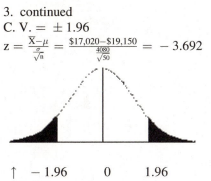

↑ − 1.96   0   1.96
− 3.692

Reject the null hypothesis. There is enough evidence to support the claim that the mean is not $19,150.

$$\overline{X} - z_{\frac{\alpha}{2}} \frac{\sigma}{\sqrt{n}} < \mu < \overline{X} + z_{\frac{\alpha}{2}} \frac{\sigma}{\sqrt{n}}$$

$$\$17,020 - 1.96\left(\frac{4080}{\sqrt{50}}\right) < \mu <$$
$$\$17,020 + 1.96\left(\frac{4080}{\sqrt{50}}\right)$$

$15,889.08 < \mu < $18,150.92$ or
$15,889 < \mu < $18,151$

The 95% confidence interval supports the conclusion because it does not contain the hypothesized mean.

5.
H_0: $\mu = 19$
H_1: $\mu \neq 19$ (claim)

C. V. = ± 2.145
$$t = \frac{\overline{X} - \mu}{\frac{s}{\sqrt{n}}} = \frac{21.3 - 19}{\frac{6.5}{\sqrt{15}}} = 1.37$$

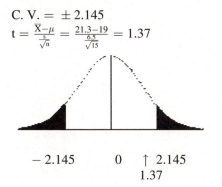

− 2.145   0   ↑ 2.145
1.37

The 99% confidence interval of the mean is:

$$\overline{X} - z_{\frac{\alpha}{2}} \frac{\sigma}{\sqrt{n}} < \mu < \overline{X} + z_{\frac{\alpha}{2}} \frac{\sigma}{\sqrt{n}}$$

$$21.3 - 2.145 \cdot \frac{6.5}{\sqrt{15}} < \mu < 21.3 + 2.145 \cdot \frac{6.5}{\sqrt{15}}$$

$17.7 < \mu < 24.9$

5. continued
The decision is do not reject the null hypothesis since $1.37 < 2.145$ and the 99% confidence interval does contain the hypothesized mean of 19. The conclusion is that there is not enough evidence to support the claim that the average time worked at home is not 19 hours per week.

7.
The power of a statistical test is the probability of rejecting the null hypothesis when it is false.

9.
The power of a test can be increased by increasing α or selecting a larger sample size.

REVIEW EXERCISES - CHAPTER 8

1.
H_0: $\mu = 98°$ (claim)
H_1: $\mu \neq 98°$

C. V. = ± 1.96
$$z = \frac{\overline{X} - \mu}{\frac{s}{\sqrt{n}}} = \frac{95.8 - 98}{\frac{7.71}{\sqrt{50}}} = -2.02$$

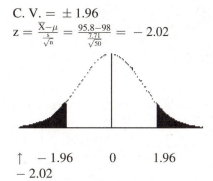

↑ − 1.96   0   1.96
− 2.02

Reject the null hypothesis. There is enough evidence to reject the claim that the average high temperature is 98°.

3.
H_0: $\mu = 18,000$
H_1: $\mu < 18,000$ (claim)

$\overline{X} = 16,298.37$ $s = 2604.82$

C. V. = − 2.326
$$z = \frac{\overline{X} - \mu}{\frac{s}{\sqrt{n}}} = \frac{16,298.37 - 18,000}{\frac{2604.82}{\sqrt{30}}} = -3.578$$

3. continued

↑ − 2.326   0
− 3.578

Reject the null hypothesis. There is enough evidence to support the claim that average debt is less than \$18,000.

5.
H_0: $\mu = \$1229$
H_1: $\mu \neq \$1229$ (claim)

C. V. = ± 1.96
$t = \frac{\overline{X} - \mu}{\frac{s}{\sqrt{n}}} = \frac{\$1350 - \$1229}{\frac{\$250}{\sqrt{15}}} = 1.875$

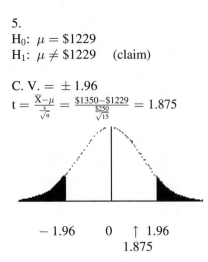

− 1.96   0   ↑ 1.96
1.875

Do not reject the null hypothesis. There is not enough evidence to support the claim that the average rent differs from \$1229.

7.
H_0: $\mu = 10$
H_1: $\mu < 10$ (claim)

C. V. = − 1.782 $\overline{X} = 9.6385$ s = 0.5853
$t = \frac{\overline{X} - \mu}{\frac{s}{\sqrt{n}}} = \frac{9.6385 - 10}{\frac{0.5853}{\sqrt{13}}} = -2.227$ or -2.230

↑ − 1.782   0
− 2.230

Reject the null hypothesis. There is enough evidence to support the claim that average weight is less than 10 ounces.

9.
H_0: $p = 0.137$
H_1: $p \neq 0.137$ (claim)

C. V. = ± 1.96
$\hat{p} = \frac{50}{300} = 0.167$ p = 0.137   q = 0.863

$z = \frac{\hat{p} - p}{\sqrt{\frac{pq}{n}}} = \frac{0.167 - 0.137}{\sqrt{\frac{(0.137)(0.863)}{300}}} = 1.51$

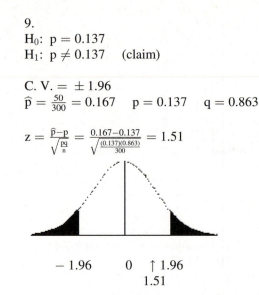

− 1.96   0   ↑ 1.96
1.51

Do not reject the null hypothesis. There is not enough evidence to support the claim that the percentage of union workers differs from 13.7%.

11.
H_0: $p = 0.593$
H_1: $p < 0.593$ (claim)

C. V. = − 2.33
$\hat{p} = \frac{156}{300} = 0.52$ p = 0.593   q = 0.407

$z = \frac{\hat{p} - p}{\sqrt{\frac{pq}{n}}} = \frac{0.52 - 0.593}{\sqrt{\frac{(0.593)(0.407)}{300}}} = -2.57$

↑ − 2.33   0
− 2.57

Reject the null hypothesis. There is enough evidence to support the claim that less than 59.3% of school lunches are free or at a reduced price.

13.
H_0: $p = 0.204$
H_1: $p \neq 0.204$ (claim)

$\hat{p} = 0.18$ p = 0.204   q = 0.796
C. V. = ± 1.96

13. continued

$$z = \frac{\hat{p}-p}{\sqrt{\frac{pq}{n}}} = \frac{0.18-0.204}{\sqrt{\frac{(0.204)(0.796)}{300}}} = -1.032$$

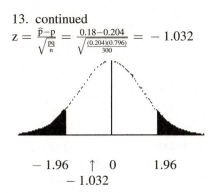

$$-1.96 \qquad \uparrow \quad 0 \qquad 1.96$$
$$-1.032$$

Do not reject the null hypothesis. There is not enough evidence to support the claim that the proportion of high school smokers differs from 20.4%.

15.
H_0: $\sigma = 4.3$ (claim)
H_1: $\sigma < 4.3$

d. f. $= 19$
$$\chi^2 = \frac{(n-1)s^2}{\sigma^2} = \frac{(20-1)(2.6)^2}{(4.3^2} = 6.95$$

$0.005 < $ P-value < 0.01 (0.006)
Since P-value < 0.05, reject the null hypothesis. There is enough evidence to reject the claim that the standard deviation is greater than or equal to 4.3 miles per gallon.

17.
H_0: $\sigma^2 = 40$
H_1: $\sigma^2 \neq 40$ (claim)

$s^2 = 43$
C. V. $= 2.700, 19.023$ d. f. $= 9$
$$\chi^2 = \frac{(n-1)s^2}{\sigma^2} = \frac{(10-1)(43)}{40} = 9.68$$

$$2.700 \qquad \uparrow \qquad\qquad 19.023$$
$$9.68$$

Do not reject the null hypothesis. There is not enough evidence to support the claim that the variance is 40.

19.
H_0: $\mu = 4$
H_1: $\mu \neq 4$ (claim)

19. continued
C. V. $= \pm 2.58$
$$z = \frac{\overline{X}-\mu}{\frac{s}{\sqrt{n}}} = \frac{4.2-4}{\frac{0.6}{\sqrt{20}}} = 1.49$$

The 99% confidence interval of the mean is:
$$\overline{X} - z_{\frac{\alpha}{2}}\frac{\sigma}{\sqrt{n}} < \mu < \overline{X} + z_{\frac{\alpha}{2}}\frac{\sigma}{\sqrt{n}}$$

$$4.2 - 2.58 \cdot \frac{0.6}{\sqrt{20}} < \mu < 4.2 + 2.58 \cdot \frac{0.6}{\sqrt{20}}$$
$$3.85 < \mu < 4.55$$

The decision is do not reject the null hypothesis since $1.49 < 2.58$ and the confidence interval does contain the hypothesized mean of 4. There is not enough evidence to support the claim that the growth has changed.

CHAPTER 8 QUIZ

1. True
2. True
3. False, the critical value separates the critical region from the noncritical region.
4. True
5. False, it can be one-tailed or two-tailed.
6. b
7. d
8. c
9. b
10. Type I
11. β
12. Statistical hypothesis
13. Right
14. $n - 1$

15. H_0: $\mu = 28.6$ (claim)
H_1: $\mu \neq 28.6$
C. V. $= \pm 1.96$
$z = 2.15$
Reject the null hypothesis. There is enough evidence to reject the claim that the average age is 28.6.

16. H_0: $\mu = \$6,500$ (claim)
H_1: $\mu \neq \$6,500$
C. V. $= \pm 1.96$
$z = 5.27$
Reject the null hypothesis. There is enough evidence to reject the agent's claim.

17. H_0: $\mu = 8$
H_1: $\mu > 8$ (claim)
C. V. $= 1.65$

17. continued
$z = 6.00$
Reject the null hypothesis. There is enough evidence to support the claim that the average is greater than 8.

18. H_0: $\mu = 500$ (claim)
H_1: $\mu \neq 500$
C. V. $= \pm 3.707$
$t = -0.571$
Do not reject the null hypothesis. There is not enough evidence to reject the claim that the average is 500.

19. H_0: $\mu = 67$
H_1: $\mu < 67$ (claim)
$t = -3.1568$
P-value < 0.005 (0.003)
Since P-value < 0.05, reject the null hypothesis. There is enough evidence to support the claim that the average height is less than 67 inches.

20. H_0: $\mu = 12.4$
H_1: $\mu < 12.4$ (claim)
C. V. $= -1.345$
$t = -2.324$
Reject the null hypothesis. There is enough evidence to support the claim that the average is less than what the company claimed.

21. H_0: $\mu = 63.5$
H_1: $\mu > 63.5$ (claim)
$t = 0.47075$
P-value > 0.25 (0.322)
Since P-value > 0.05, do not reject the null hypothesis. There is not enough evidence to support the claim that the average is greater than 63.5.

22. H_0: $\mu = 26$ (claim)
H_1: $\mu \neq 26$
C. V. $= \pm 2.492$
$t = -1.5$
Do not reject the null hypothesis. There is not enough evidence to reject the claim that the average age is 26.

23. H_0: $p = 0.39$ (claim)
H_1: $p \neq 0.39$
C. V. $= \pm 1.96$
$z = -0.62$
Do not reject the null hypothesis. There is not enough evidence to reject the claim that

23. continued
39% took supplements. The study supports the results of the previous study.

24. H_0: $p = 0.55$ (claim)
H_1: $p < 0.55$
C. V. $= -1.28$
$z = -0.899$
Do not reject the null hypothesis. There is not enough evidence to reject the dietitian's claim.

25. H_0: $p = 0.35$ (claim)
H_1: $p \neq 0.35$
C. V. $= \pm 2.33$
$z = 0.666$
Do not reject the null hypothesis. There is not enough evidence to reject the claim that the proportion is 35%.

26. H_0: $p = 0.75$ (claim)
H_1: $p \neq 0.75$
C. V. $= \pm 2.58$
$z = 2.6833$
Reject the null hypothesis. there is enough evidence to reject the claim.

27. The area corresponding to $z = 2.14$ is 0.9838.
P-value $= 2(1 - 0.9838) = 0.0324$

28. The area corresponding to $z = 5.27$ is greater than 0.9999.
Thus, P-value $\leq 2(1 - 0.9999) \leq 0.0002$.
(TI: P-value < 0.0001)

29. H_0: $\sigma = 6$
H_1: $\sigma > 6$ (claim)
C. V. $= 36.415$
$\chi^2 = 54$
Reject the null hypothesis. There is enough evidence to support the claim that the standard deviation is more than 6 pages.

30. H_0: $\sigma = 8$ (claim)
H_1: $\sigma \neq 8$
C. V. $= 27.991, 79.490$
$\chi^2 = 33.2$
Do not reject the null hypothesis. There is not enough evidence to reject the claim that $\sigma = 8$.

31. H_0: $\sigma = 2.3$
H_1: $\sigma < 2.3$ (claim)
C. V. $= 10.117$

31. continued

$\chi^2 = 13$

Do not reject the null hypothesis. There is not enough evidence to support the claim that the standard deviation is less than 2.3.

32. H_0: $\sigma = 9$ (claim)

H_1: $\sigma \neq 9$

$\chi^2 = 13.4$

P-value > 0.20 (0.290)

Since P-value > 0.05, do not reject the null hypothesis. There is not enough evidence to reject the claim that $\sigma = 9$.

33. $28.9 < \mu < 31.2$; no

34. $\$6562.81 < \mu < \$6,637.19$; no

Note: Graphs are not to scale and are intended to convey a general idea. Answers may vary due to rounding, TI-83's, or computer programs.

EXERCISE SET 9-1

1.
Testing a single mean involves comparing a sample mean to a specific value such as $\mu = 100$; whereas testing the difference between means means comparing the means of two samples such as $\mu_1 = \mu_2$.

3.
The populations must be independent of each other and they must be normally distributed. s_1 and s_2 can be used in place of σ_1 and σ_2 when σ_1 and σ_2 are unknown and both samples are each greater than or equal to 30.

5.
H_0: $\mu_1 = \mu_2$ (claim)
H_1: $\mu_1 \neq \mu_2$

C. V. $= \pm 2.58$
$\overline{X}_1 = 662.6111$ $\overline{X}_2 = 758.875$

$$z = \frac{(\overline{X}_1 - \overline{X}_2) - (\mu_1 - \mu_2)}{\sqrt{\frac{\sigma_1^2}{n_1} + \frac{\sigma_2^2}{n_2}}} = \frac{(662.6111 - 758.875) - 0}{\sqrt{\frac{450^2}{36} + \frac{474^2}{36}}} =$$

$z = -0.88$
(TI83 answer is $z = -0.856$)

-2.58 ↑ 0 2.58
-0.88

Do not reject the null hypothesis. There is not enough evidence to reject the claim that the average lengths of the rivers is the same.

7.
H_0: $\mu_1 = \mu_2$
H_1: $\mu_1 \neq \mu_2$ (claim)

C. V. $= \pm 1.96$

7. continued
$$z = \frac{(\overline{X}_1 - \overline{X}_2) - (\mu_1 - \mu_2)}{\sqrt{\frac{\sigma_1^2}{n_1} + \frac{\sigma_2^2}{n_2}}} = \frac{(28.5 - 35.2) - 0}{\sqrt{\frac{7.2^2}{40} + \frac{9.1^2}{40}}} = -3.65$$

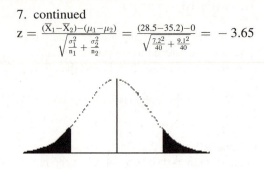

↑ -1.96 0 1.96
-3.65

Reject the null hypothesis. There is enough evidence to support the claim that commuting times are different in the winter.

9.
H_0: $\mu_1 = \mu_2$
H_1: $\mu_1 > \mu_2$ (claim)

C. V. $= 2.33$

$$z = \frac{(\overline{X}_1 - \overline{X}_2) - (\mu_1 - \mu_2)}{\sqrt{\frac{\sigma_1^2}{n_1} + \frac{\sigma_2^2}{n_2}}} = \frac{(5.5 - 4.2) - 0}{\sqrt{\frac{1.2^2}{32} + \frac{1.5^2}{30}}} = 3.75$$

0 2.33 ↑
3.75

Reject the null hypothesis. There is enough evidence to support the claim that the average stay is longer for men than for women.

11.
H_0: $\mu_1 = \mu_2$
H_1: $\mu_1 < \mu_2$ (claim)

C. V. $= -1.65$

$$z = \frac{(\overline{X}_1 - \overline{X}_2) - (\mu_1 - \mu_2)}{\sqrt{\frac{\sigma_1^2}{n_1} + \frac{\sigma_2^2}{n_2}}} = \frac{(3.16 - 3.28) - 0}{\sqrt{\frac{0.52^2}{103} + \frac{0.46^2}{225}}} = -2.01$$

11. continued

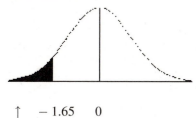

$\uparrow \quad -1.65 \quad 0$
-2.01

Reject the null hypothesis. There is enough
evidence to support the claim that those who
stayed have a higher GPA than those who
left their profession.

13.
H_0: $\mu_1 = \mu_2$
H_1: $\mu_1 \neq \mu_2$ (claim)

C. V. = ± 1.96

$z = \dfrac{(\overline{X}_1 - \overline{X}_2) - (\mu_1 - \mu_2)}{\sqrt{\frac{\sigma_1^2}{n_1} + \frac{\sigma s_2^2}{n_2}}}$

$z = \dfrac{(40{,}275 - 38{,}750) - 0}{\sqrt{\frac{10{,}500^2}{50} + \frac{12{,}500^2}{50}}} = 0.66$

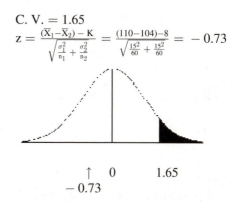

$-1.96 \qquad 0 \quad \uparrow \ 1.96$
0.66

Do not reject the null hypothesis. There is
not enough evidence to support the claim
that the average incomes are different.

15.
H_0: $\mu_1 = \mu_2$
H_1: $\mu_1 \neq \mu_2$ (claim)

$z = \dfrac{(\overline{X}_1 - \overline{X}_2) - (\mu_1 - \mu_2)}{\sqrt{\frac{\sigma_1^2}{n_1} + \frac{\sigma_2^2}{n_2}}} = \dfrac{(3.05 - 2.96) - 0}{\sqrt{\frac{0.75^2}{103} + \frac{0.75^2}{225}}}$

$z = 1.01$
Area $= 0.8438$
P-value $= 2(1 - 0.8438) = 0.3124$
Since P-value > 0.05, do not reject the null
hypothesis. There is not enough evidence to
support the claim that there is a difference in
scores. (TI: P-value $= 0.3131$)

17.
$\overline{D} = 83.6 - 79.2 = 4.4$
$(\overline{X}_1 - \overline{X}_2) - z_{\frac{\alpha}{2}} \sqrt{\frac{\sigma_1^2}{n_1} + \frac{\sigma_2^2}{n_2}} < \mu_1 - \mu_2 <$
$\qquad (\overline{X}_1 - \overline{X}_2) + z_{\frac{\alpha}{2}} \sqrt{\frac{\sigma_1^2}{n_1} + \frac{\sigma_2^2}{n_2}}$

$4.4 - (1.65)\sqrt{\frac{4.3^2}{36} + \frac{3.8^2}{36}} < \mu_1 - \mu_2 <$
$\qquad 4.4 + (1.65)\sqrt{\frac{4.3^2}{36} + \frac{3.8^2}{36}}$

$2.8 < \mu_1 - \mu_2 < 6.0$
(TI: $2.83 < \mu_1 - \mu_2 < 5.97$)

19.
$\overline{D} = 315 - 280 = 35$
$(\overline{X}_1 - \overline{X}_2) - z_{\frac{\alpha}{2}} \sqrt{\frac{\sigma_1^2}{n_1} + \frac{\sigma_2^2}{n_2}} < \mu_1 - \mu_2 <$
$\qquad (\overline{X}_1 - \overline{X}_2) + z_{\frac{\alpha}{2}} \sqrt{\frac{\sigma_1^2}{n_1} + \frac{\sigma_2^2}{n_2}}$

$35 - (1.96)\sqrt{\frac{56.2^2}{40} + \frac{52.1^2}{35}} < \mu_1 - \mu_2 <$

$\qquad 35 + (1.96)\sqrt{\frac{56.2^2}{40} + \frac{52.1^2}{35}}$

$10.48 < \mu_1 - \mu_2 < 59.52$

The interval gives evidence to reject the
claim that there is no difference in the means
because 0 is not contained in the interval.

21.
H_0: $\mu_1 - \mu_2 = 8$ (claim)
H_1: $\mu_1 - \mu_2 > 8$

C. V. = 1.65
$z = \dfrac{(\overline{X}_1 - \overline{X}_2) - K}{\sqrt{\frac{\sigma_1^2}{n_1} + \frac{\sigma_2^2}{n_2}}} = \dfrac{(110 - 104) - 8}{\sqrt{\frac{15^2}{60} + \frac{15^2}{60}}} = -0.73$

$\uparrow \quad 0 \qquad 1.65$
-0.73

Do not reject the null hypothesis. There is
not enough evidence to reject the claim that
private school students have exam scores
that are at most 8 points higher than public
school students.

23.
H_0: $\mu_1 - \mu_2 = \$30,000$
H_1: $\mu_1 - \mu_2 \neq \$30,000$ (claim)

C. V. = ± 2.58

$$z = \frac{(\overline{X}_1 - \overline{X}_2) - K}{\sqrt{\frac{\sigma_1^2}{n_1} + \frac{\sigma_2^2}{n_2}}}$$

$$z = \frac{(90,200 - 57,800) - 30,000}{\sqrt{\frac{15,000^2}{100} + \frac{12,800^2}{100}}} = 1.22$$

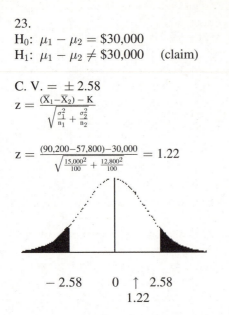

-2.58 0 ↑ 2.58
                 1.22

Do not reject the null hypothesis. There is not enough evidence to support the claim that the difference in income is not $30,000.

EXERCISE SET 9-2

1.
H_0: $\mu_1 = \mu_2$
H_1: $\mu_1 \neq \mu_2$ (claim)

C. V. = ± 1.761 d. f. = 14

$$t = \frac{(\overline{X}_1 - \overline{X}_2) - (\mu_1 - \mu_2)}{\sqrt{\frac{s_1^2}{n_1} + \frac{s_2^2}{n_2}}}$$

$$t = \frac{(22 - 28) - 0}{\sqrt{\frac{6.17^2}{15} + \frac{13.2^2}{15}}} = -1.595$$

-1.761 ↑   0   1.761
       -1.595

Do not reject the null hypothesis. There is not enough evidence to support the claim that there is a significant difference in the average number of weeks on the bestseller lists.

3.
H_0: $\mu_1 = \mu_2$
H_1: $\mu_1 \neq \mu_2$ (claim)

3. continued
C. V. = ± 2.093 d. f. = 19

$$t = \frac{(\overline{X}_1 - \overline{X}_2) - (\mu_1 - \mu_2)}{\sqrt{\frac{s_1^2}{n_1} + \frac{s_2^2}{n_2}}}$$

$$t = \frac{(63.1 - 56.3) - 0}{\sqrt{\frac{4.1^2}{20} + \frac{7.5^2}{24}}} = 3.811$$

-2.093 0   2.093   ↑
                                        3.811

Reject the null hypothesis. There is enough evidence to support the claim that there is a difference in noise levels.

5.
H_0: $\mu_1 = \mu_2$
H_1: $\mu_1 \neq \mu_2$ (claim)

C. V. = ± 1.812 d. f. = 10
$\overline{X}_1 = 29.69$ $s_1 = 6.499$
$\overline{X}_2 = 34.36$ $s_2 = 11.201$

$$t = \frac{(\overline{X}_1 - \overline{X}_2) - (\mu_1 - \mu_2)}{\sqrt{\frac{s_1^2}{n_1} + \frac{s_2^2}{n_2}}}$$

$$t = \frac{(29.69 - 34.36) - 0}{\sqrt{\frac{6.499^2}{13} + \frac{11.201^2}{11}}} = -1.220$$

-1.812 ↑   0   1.812
    -1.220

Do not reject the null hypothesis. There is not enough evidence to support the claim that there is a significant difference in the average number of grams of carbohydrates.

7.
H_0: $\mu_1 = \mu_2$
H_1: $\mu_1 \neq \mu_2$ (claim)

$\overline{X}_1 = 12.48$ $s_1 = 1.477$
$\overline{X}_2 = 9.94$ $s_2 = 0.666$

7. continued

$$t = \frac{(\overline{X}_1 - \overline{X}_2) - (\mu_1 - \mu_2)}{\sqrt{\frac{s_1^2}{n_1} + \frac{s_2^2}{n_2}}}$$

$$t = \frac{(12.48 - 9.94) - 0}{\sqrt{\frac{1.477^2}{10} + \frac{0.666^2}{15}}} = 5.10$$

For z = 5.10, the area is greater than 0.9999. The P-value is $1 - 0.9999 < 0.0001$. Since the P-value is less than $\alpha = 0.05$, reject the null hypothesis. There is enough evidence to support the claim that there is a difference in the weights of running shoes.

9.
For the 95% Confidence Interval:
$6.8 - 2.093(1.784) < \mu_1 - \mu_2 <$
$\qquad 6.8 + 2.093(1.784)$

$3.066 < \mu_1 - \mu_2 < 10.534$
(TI:$3.18 < \mu_1 - \mu_2 < 10.42$)

11.
$H_0: \mu_1 = \mu_2$
$H_1: \mu_1 \neq \mu_2$ (claim)

C. V. = ± 2.977 d. f. = 14
$t = \frac{(22.45 - 18.5) - 0}{\sqrt{\frac{16.4}{15} + \frac{18.2}{15}}} = 2.60$

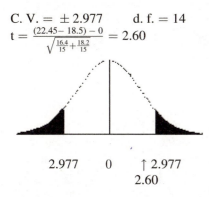

2.977   0   ↑ 2.977
                        2.60

Do not reject the null hypothesis. There is not enough evidence to support the claim that there is a difference in TV viewing times between children and teens.

13.
$H_0: \mu_1 = \mu_2$
$H_1: \mu_1 > \mu_2$ (claim)

$\overline{X}_1 = 39.6667$ $s_1 = 18.3703$
$\overline{X}_2 = 28.8333$ $s_2 = 17.1279$

C. V. = 3.365   d. f. = 5
$t = \frac{(39.6667 - 28.8333) - 0}{\sqrt{\frac{18.3703^2}{6} + \frac{17.1279^2}{6}}} = 1.057$

13. continued

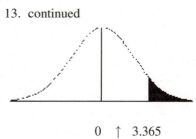

0   ↑   3.365
    1.057

Do not reject the null hypothesis. There is not enough evidence to support the claim that the average number of students attending cyber schools in Allegheny County is greater than those who attend cyber schools outside Allegheny County. One reason why caution should be used is that cyber charter schools are a relatively new concept.

15.
$H_0: \mu_1 = \mu_2$ (claim)
$H_1: \mu_1 \neq \mu_2$

P-value: $0.02 < $ P-value < 0.05 (0.026)
d. f. = 15
$t = \frac{(2.3 - 1.9) - 0}{\sqrt{\frac{0.6^2}{16} + \frac{0.3^2}{16}}} = 2.385$

Since P-value > 0.01, do not reject the null hypothesis. There is not enough evidence to reject the claim that the mean hospital stay is the same. (TI: P-value = 0.026)

99% Confidence Interval:
$0.4 - 2.947(0.1677) < \mu_1 - \mu_2 <$
$\qquad 0.4 + 2.947(0.1677)$

$-0.09 < \mu_1 - \mu_2 < 0.89$
(TI: $-0.07 < \mu_1 - \mu_2 < 0.87$)

17.
Research: $\overline{X}_1 = 596.2353$ $s_1 = 163.2362$
Primary Care: $\overline{X}_2 = 481.5$ $s_2 = 179.3957$
90% Confidence Interval:

$114.7353 - 1.753\sqrt{\frac{163.2363^2}{17} + \frac{179.3957^2}{16}}$
$\qquad < \mu_1 - \mu_2 <$
$114.7353 + 1.753\sqrt{\frac{163.2363^2}{17} + \frac{179.3957^2}{16}}$

$114.7353 - 104.87 < \mu_1 - \mu_2$
$\qquad < 114.7353 + 104.87$
$9.8653 < \mu_1 - \mu_2 < 219.6053$
(TI: $13.2 < \mu_1 - \mu_2 < 216.2$)

EXERCISE SET 9-3

1.
a. Dependent
b. Dependent
c. Independent
d. Dependent
e. Independent

3.

Before	After	D	D^2
9	9	0	0
12	17	-5	25
6	9	-3	9
15	20	-5	25
3	2	1	1
18	21	-3	9
10	15	-5	25
13	22	-9	81
7	6	1	1
		$\sum D = -28$	$\sum D^2 = 176$

H_0: $\mu_D = 0$
H_1: $\mu_D < 0$ (claim)

C. V. $= -1.397$ d. f. $= 8$
$\overline{D} = \frac{\sum D}{n} = -3.11$

$s_D = \sqrt{\frac{n\sum D^2 - (\sum D)^2}{n(n-1)}} = \sqrt{\frac{9(176) - (-28)^2}{9(8)}} = 3.3$

$t = \frac{-3.11 - 0}{\frac{3.33}{\sqrt{9}}} = -2.8$

$\uparrow -1.397$ 0
-2.8

Reject the null hypothesis. There is enough evidence to support the claim that the seminar increased the number of hours students studied.

5.

F - S	S - Th	D	D^2
4	8	-4	16
7	5.5	1.5	2.25
10.5	7.5	3	9
12	8	4	16
11	7	4	16
9	6	3	9
6	6	0	0
9	8	1	1
		$\sum D = 12.5$	$\sum D^2 = 69.25$

H_0: $\mu_D = 0$
H_1: $\mu_D \neq 0$ (claim)

C. V. $= \pm 2.365$ d. f. $= 7$
$\overline{D} = \frac{\sum D}{n} = \frac{12.5}{8} = 1.5625$

$s_D = \sqrt{\frac{n\sum D^2 - (\sum D)^2}{n(n-1)}}$

$= \sqrt{\frac{8(69.25) - (12.5)^2}{8(7)}} = 2.665$

$t = \frac{1.5625 - 0}{\frac{2.665}{\sqrt{8}}} = 1.6583$

-2.365 0 $\uparrow$ 2.365
                   1.6583

Do not reject the null hypothesis. There is not enough evidence to support the claim that there is a difference in the mean number of hours slept.

7.

Before	After	D	D^2
12	9	3	9
9	6	3	9
0	1	-1	1
5	3	2	4
4	2	2	4
3	3	0	0
		$\sum D = 9$	$\sum D^2 = 27$

H_0: $\mu_D = 0$
H_1: $\mu_D > 0$ (claim)

C. V. $= 2.571$ d. f. $= 5$
$\overline{D} = \frac{\sum D}{n} = \frac{9}{6} = 1.5$

7. continued

$s_D = \sqrt{\frac{n\sum D^2 - (\sum D)^2}{n(n-1)}} = \sqrt{\frac{6(27) - 9^2}{6(5)}} = 1.64$

$t = \frac{1.5 - 0}{\frac{1.64}{\sqrt{6}}} = 2.24$

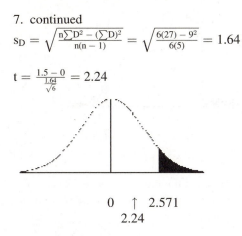

```
0    ↑   2.571
        2.24
```

Do not reject the null hypothesis. There is not enough evidence to support the claim that the errors have been reduced.

9.

A	B	D	D²
87	83	4	16
92	95	-3	9
78	79	-1	1
83	83	0	0
88	86	2	4
90	93	-3	9
84	80	4	16
93	86	7	49

$\sum D = 10 \quad \sum D^2 = 104$

$H_0: \mu_D = 0$
$H_1: \mu_D \neq 0$ (claim)

P-value = 0.361   d. f. = 7
$\overline{D} = \frac{\sum D}{n} = \frac{10}{8} = 1.25$

$s_D = \sqrt{\frac{n\sum D^2 - (\sum D)^2}{n(n-1)}} = \sqrt{\frac{8(104) - 10^2}{8(7)}} = 3.62$

$t = \frac{1.25 - 0}{\frac{3.62}{\sqrt{8}}} = 0.978$

0.20 < P-value < 0.50 Do not reject the null hypothesis since P-value > 0.01. There is not enough evidence to support the claim that there is a difference in the pulse rates.

Confidence Interval:
$1.25 - 3.499\left(\frac{3.62}{\sqrt{8}}\right) < \mu_D <$
$\qquad 1.25 + 3.499\left(\frac{3.62}{\sqrt{8}}\right)$
$-3.23 < \mu_D < 5.73$

11.
$\overline{X}_1 - \overline{X}_2 = \sum \frac{X_1 - X_2}{n}$

11. continued

$\sum \frac{X_1 - X_2}{n} = \sum \left(\frac{X_1}{n} - \frac{X_2}{n}\right)$

$\sum \left(\frac{X_1}{n} - \frac{X_2}{n}\right) = \sum \frac{X_1}{n} - \sum \frac{X_2}{n}$

$\sum \frac{X_1}{n} - \sum \frac{X_2}{n} = \overline{X}_1 - \overline{X}_2$

EXERCISE SET 9-4

1A.
Use $\hat{p} = \frac{X}{n}$ and $\hat{q} = 1 - \hat{p}$

a. $\hat{p} = \frac{34}{48}$ $\qquad \hat{q} = \frac{14}{48}$

b. $\hat{p} = \frac{28}{75}$ $\qquad \hat{q} = \frac{47}{75}$

c. $\hat{p} = \frac{50}{100}$ $\qquad \hat{q} = \frac{50}{100}$

d. $\hat{p} = \frac{6}{24}$ $\qquad \hat{q} = \frac{18}{24}$

e. $\hat{p} = \frac{12}{144}$ $\qquad \hat{q} = \frac{132}{144}$

1B.
a. x = 0.16(100) = 16
b. x = 0.08(50) = 4
c. x = 0.06(800) = 48
d. x = 0.52(200) = 104
e. x = 0.20(150) = 30

3.
$\hat{p}_1 = \frac{178}{300} = 0.593 \quad \hat{p}_2 = \frac{139}{300} = 0.463$

$\overline{p} = \frac{X_1 + X_2}{n_1 + n_2} = \frac{178 + 139}{300 + 300} = \frac{317}{600} = 0.528$

$\overline{q} = 1 - \overline{p} = 1 - 0.5283 = 0.472$

$H_0: p_1 = p_2$
$H_1: p_1 > p_2$ (claim)

C. V. = 1.645
$z = \frac{(\hat{p}_1 - \hat{p}_2) - (p_1 - p_2)}{\sqrt{(\overline{p})(\overline{q})\left(\frac{1}{n_1} + \frac{1}{n_2}\right)}} = \frac{(0.593 - 0.463) - 0}{\sqrt{(0.528)(0.472)\left(\frac{1}{300} + \frac{1}{300}\right)}}$

$z = 3.19$

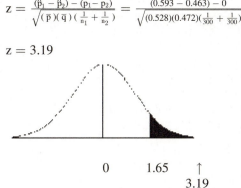

```
0       1.65      ↑
                 3.19
```

3. continued

Reject the null hypothesis. There is enough evidence to support the claim that the proportion of married men is greater than the proportion of married women.

5.

$\hat{p}_1 = \frac{X_1}{n_1} = \frac{980}{1200} = 0.8167$

$\hat{p}_2 = \frac{940}{1200} = 0.7833$

$\bar{p} = \frac{X_1 + X_2}{n_1 + n_2} = \frac{980 + 940}{1200 + 1200} = 0.8$

$\bar{q} = 1 - \bar{p} = 1 - 0.8 = 0.2$

H_0: $p_1 = p_2$
H_1: $p_1 \neq p_2$ (claim)

C. V. $= \pm 1.96$

$z = \frac{(\hat{p}_1 - \hat{p}_2) - (p_1 - p_2)}{\sqrt{(\bar{p})(\bar{q})(\frac{1}{n_1} + \frac{1}{n_2})}} = \frac{(0.8167 - 0.7833) - 0}{\sqrt{(0.8)(0.2)(\frac{1}{1200} + \frac{1}{1200})}}$

$z = 2.045$
(TI = 2.041)

$-1.96 \qquad 0 \qquad 1.96 \; \uparrow$
$\qquad\qquad\qquad\qquad 2.045$

Reject the null hypothesis. There is enough evidence to support the claim that the proportions are different.

7.

$\hat{p}_1 = 0.83 \qquad \hat{p}_2 = 0.75$
$X_1 = 0.83(100) = 83$
$X_2 = 0.75(100) = 75$

$\bar{p} = \frac{83 + 75}{100 + 100} = 0.79 \qquad \bar{q} = 1 - 0.79 = 0.21$

H_0: $p_1 = p_2$ (claim)
H_1: $p_1 \neq p_2$

C. V. $= \pm 1.96 \qquad \alpha = 0.05$

$z = \frac{(\hat{p}_1 - \hat{p}_2) - (p_1 - p_2)}{\sqrt{(\bar{p})(\bar{q})(\frac{1}{n_1} + \frac{1}{n_2})}} = \frac{(0.83 - 0.75) - 0}{\sqrt{(0.79)(0.21)(\frac{1}{100} + \frac{1}{100})}}$

$z = 1.39$

7. continued

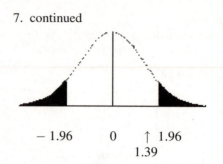

$-1.96 \qquad 0 \qquad \uparrow 1.96$
$\qquad\qquad\qquad 1.39$

Do not reject the null hypothesis. There is not enough evidence to reject the claim that the proportions are equal.

$(\hat{p}_1 - \hat{p}_2) - z_{\frac{\alpha}{2}} \sqrt{\frac{\hat{p}_1 \hat{q}_1}{n_1} + \frac{\hat{p}_2 \hat{q}_2}{n_2}} < p_1 - p_2 <$

$\qquad (\hat{p}_1 - \hat{p}_2) + z_{\frac{\alpha}{2}} \sqrt{\frac{\hat{p}_1 \hat{q}_1}{n_1} + \frac{\hat{p}_2 \hat{q}_2}{n_2}}$

$0.08 - 1.96 \sqrt{\frac{0.83(0.17)}{100} + \frac{0.75(0.25)}{100}} < p_1 - p_2$

$\qquad < 0.08 + 1.96 \sqrt{\frac{0.83(0.17)}{100} + \frac{0.75(0.25)}{100}}$

$-0.032 < p_1 - p_2 < 0.192$

9.

$\hat{p}_1 = \frac{44}{80} = 0.55 \qquad \hat{p}_2 = \frac{41}{90} = 0.4556$

$\bar{p} = \frac{X_1 + X_2}{n_1 + n_2} = \frac{44 + 41}{80 + 90} = 0.5$

$\bar{q} = 1 - \bar{p} = 1 - 0.5 = 0.5$

H_0: $p_1 = p_2$
H_1: $p_1 \neq p_2$ (claim)

C. V. $= \pm 2.58 \qquad \alpha = 0.01$

$z = \frac{(\hat{p}_1 - \hat{p}_2) - (p_1 - p_2)}{\sqrt{(\bar{p})(\bar{q})(\frac{1}{n_1} + \frac{1}{n_2})}} = \frac{(0.55 - 0.4556) - 0}{\sqrt{(0.5)(0.5)(\frac{1}{80} + \frac{1}{90})}}$

$z = 1.23$

$-2.58 \qquad 0 \qquad \uparrow 2.58$
$\qquad\qquad\qquad 1.23$

9. continued

Do not reject the null hypothesis. There is not enough evidence to support the claim that the proportions are different.

$$(\hat{p}_1 - \hat{p}_2) - z_{\frac{\alpha}{2}}\sqrt{\frac{\hat{p}_1\hat{q}_1}{n_1} + \frac{\hat{p}_2\hat{q}_2}{n_2}} < p_1 - p_2 <$$

$$(\hat{p}_1 - \hat{p}_2) + z_{\frac{\alpha}{2}}\sqrt{\frac{\hat{p}_1\hat{q}_1}{n_1} + \frac{\hat{p}_2\hat{q}_2}{n_2}}$$

$$0.0944 - 2.58\sqrt{\frac{0.55(0.4556)}{80} + \frac{0.4556(0.55)}{90}} < p_1 - p_2$$

$$< 0.0944 + 2.58\sqrt{\frac{0.55(0.4556)}{80} + \frac{0.4556(0.55)}{90}}$$

$$-0.104 < p_1 - p_2 < 0.293$$
(TI: $-0.103 < p_1 - p_2 < 0.291$)

11.

$$\hat{p}_1 = \frac{26}{75} = 0.3467 \qquad \hat{p}_2 = \frac{26}{60} = 0.4333$$

$$\bar{p} = \frac{X_1 + X_2}{n_1 + n_2} = \frac{26 + 26}{75 + 60} = 0.3852$$

$$\bar{q} = 1 - \bar{p} = 1 - 0.3852 = 0.6148$$

H_0: $p_1 = p_2$
H_1: $p_1 \neq p_2$ (claim)

C. V. $= \pm 1.96$ $\alpha = 0.05$

$$z = \frac{(\hat{p}_1 - \hat{p}_2) - (p_1 - p_2)}{\sqrt{(\bar{p})(\bar{q})\left(\frac{1}{n_1} + \frac{1}{n_2}\right)}} = \frac{(0.3467 - 0.4333) - 0}{\sqrt{(0.3852)(0.6148)\left(\frac{1}{75} + \frac{1}{60}\right)}}$$

$$z = -1.027$$

-1.96 ↑ 0 $\quad$ 1.96
$\quad$ -1.027

Do not reject the null hypothesis. There is not enough evidence to support the claim that the proportin of dog owners has changed.

$$(\hat{p}_1 - \hat{p}_2) - z_{\frac{\alpha}{2}}\sqrt{\frac{\hat{p}_1\hat{q}_1}{n_1} + \frac{\hat{p}_2\hat{q}_2}{n_2}} < p_1 - p_2 <$$

$$(\hat{p}_1 - \hat{p}_2) + z_{\frac{\alpha}{2}}\sqrt{\frac{\hat{p}_1\hat{q}_1}{n_1} + \frac{\hat{p}_2\hat{q}_2}{n_2}}$$

11. continued

$$-0.0866 - 1.96\sqrt{\frac{0.3467(0.6533)}{75} + \frac{0.4333(0.5667)}{60}}$$
$$< p_1 - p_2 <$$

$$-0.0866 + 1.96\sqrt{\frac{0.3467(0.6533)}{75} + \frac{0.4333(0.5667)}{60}}$$

$$-0.252 < p_1 - p_2 < 0.079$$

Yes, the confidence interval contains 0. This is another way to conclude that there is no difference in the proportions.

13.

$$\hat{p}_1 = \frac{X_1}{n_1} = \frac{50}{200} = 0.25$$

$$\hat{p}_2 = \frac{X_2}{n_2} = \frac{93}{300} = 0.31$$

$$\bar{p} = \frac{X_1 + X_2}{n_1 + n_2} = \frac{50 + 93}{200 + 300} = 0.286$$

$$\bar{q} = 1 - \bar{p} = 1 - 0.286 = 0.714$$

H_0: $p_1 = p_2$
H_1: $p_1 \neq p_2$ (claim)

C. V. $= \pm 2.58$ $\alpha = 0.01$

$$z = \frac{(\hat{p}_1 - \hat{p}_2) - (p_1 - p_2)}{\sqrt{(\bar{p})(\bar{q})\left(\frac{1}{n_1} + \frac{1}{n_2}\right)}} = \frac{(0.25 - 0.31) - 0}{\sqrt{(0.286)(0.714)\left(\frac{1}{200} + \frac{1}{300}\right)}}$$

$$z = -1.45$$

-2.58 $\quad$ ↑ 0 $\quad$ 2.58
$\quad$ -1.45

Do not reject the null hypothesis. There is not enough evidence to support the claim that the proportions are different.

$$(\hat{p}_1 - \hat{p}_2) - z_{\frac{\alpha}{2}}\sqrt{\frac{\hat{p}_1\hat{q}_1}{n_1} + \frac{\hat{p}_2\hat{q}_2}{n_2}} < p_1 - p_2 <$$

$$(\hat{p}_1 - \hat{p}_2) + z_{\frac{\alpha}{2}}\sqrt{\frac{\hat{p}_1\hat{q}_1}{n_1} + \frac{\hat{p}_2\hat{q}_2}{n_2}}$$

13. continued

$$-0.06 - 2.58\sqrt{\frac{0.25(0.75)}{200} + \frac{0.31(0.69)}{300}} <$$

$$p_1 - p_2 < -0.06 + 2.58\sqrt{\frac{0.25(0.75)}{200} + \frac{0.31(0.69)}{300}}$$

$$-0.165 < p_1 - p_2 < 0.045$$

15.
$\alpha = 0.01$
$\hat{p}_1 = 0.8 \qquad \hat{q}_1 = 0.2$
$\hat{p}_2 = 0.6 \qquad \hat{q}_2 = 0.4$

$$\hat{p}_1 - \hat{p}_2 = 0.8 - 0.6 = 0.2$$

$$(\hat{p}_1 - \hat{p}_2) - z_{\frac{\alpha}{2}}\sqrt{\frac{\hat{p}_1\hat{q}_1}{n_1} + \frac{\hat{p}_2\hat{q}_2}{n_2}} < p_1 - p_2 <$$

$$(\hat{p}_1 - \hat{p}_2) + z_{\frac{\alpha}{2}}\sqrt{\frac{\hat{p}_1\hat{q}_1}{n_1} + \frac{\hat{p}_2\hat{q}_2}{n_2}}$$

$$0.2 - 2.58\sqrt{\frac{(0.8)(0.2)}{150} + \frac{(0.6)(0.4)}{200}} < p_1 - p_2 <$$

$$0.2 + 2.58\sqrt{\frac{(0.8)(0.2)}{150} + \frac{(0.6)(0.4)}{200}}$$

$$0.077 < p_1 - p_2 < 0.323$$

17.
$\hat{p}_1 = \frac{X_1}{n_1} = \frac{80}{200} = 0.4 \qquad \hat{p}_2 = \frac{59}{200} = 0.295$

$$\bar{p} = \frac{X_1 + X_2}{n_1 + n_2} = \frac{80 + 59}{200 + 200} = 0.3475$$

$$\bar{q} = 1 - \bar{p} = 1 - 0.3475 = 0.6525$$

H_0: $p_1 = p_2$
H_1: $p_1 \neq p_2$ (claim)

C. V. $= \pm 2.58$
$$z = \frac{(\hat{p}_1 - \hat{p}_2) - (p_1 - p_2)}{\sqrt{(\bar{p})(\bar{q})(\frac{1}{n_1} + \frac{1}{n_2})}} = \frac{(0.4 - 0.295) - 0}{\sqrt{(0.3475)(0.6525)(\frac{1}{200} + \frac{1}{200})}}$$

$z = 2.21$

$-2.58 \qquad 0 \qquad \uparrow 2.58$
$\qquad\qquad\qquad 2.21$

17. continued
Do not reject the null hypothesis. There is not enough evidence to support the claim that the proportions are different.

19.
$\hat{p}_1 = \frac{115}{400} = 0.2875 \qquad \hat{q}_1 = 0.7125$
$\hat{p}_2 = \frac{100}{350} = 0.2857 \qquad \hat{q}_2 = 0.7143$
$\hat{p}_1 - \hat{p}_2 = 0.0018$

$$(\hat{p}_1 - \hat{p}_2) - z_{\frac{\alpha}{2}}\sqrt{\frac{\hat{p}_1\hat{q}_1}{n_1} + \frac{\hat{p}_2\hat{q}_2}{n_2}} < p_1 - p_2 <$$

$$(\hat{p}_1 - \hat{p}_2) + z_{\frac{\alpha}{2}}\sqrt{\frac{\hat{p}_1\hat{q}_1}{n_1} + \frac{\hat{p}_2\hat{q}_2}{n_2}}$$

$$0.0018 - 1.96\sqrt{\frac{(0.2875)(0.7125)}{400} + \frac{(0.2857)(0.7143)}{350}} < p_1 - p_2$$

$$< 0.0018 + 1.96\sqrt{\frac{(0.2875)(0.7125)}{400} + \frac{(0.2857)(0.7143)}{350}}$$

$$-0.0631 < p_1 - p_2 < 0.0667$$

The interval does agree with the *Almanac* statistics stating a difference of -0.042 since -0.042 is contained in the interval.

EXERCISE SET 9-5

1.
It should be the larger of the two variances.

3.
One d.f. is used for the variance associated with the numerator and one is used for the variance associated with the denominator.

5.
a. d. f. N = 15, d. f. D = 22; C. V. = 3.36
b. d. f. N = 24, d. f. D = 13; C. V. = 3.59
c. d. f. N = 45, d. f. D = 29; C. V. = 2.03
d. d. f. N = 20, d. f. D = 16; C. V. = 2.28
e. d. f. N = 10, d. f. D = 10; C. V. = 2.98

6.
Note: Specific P-values are in parentheses.
a. $0.025 <$ P-value < 0.05 (0.033)
b. $0.05 <$ P-value < 0.10 (0.072)
c. P-value $= 0.05$
d. $0.005 <$ P-value < 0.01 (0.006)
e. P-value $= 0.05$
f. P > 0.10 (0.112)
g. $0.05 <$ P-value < 0.10 (0.068)
h. $0.01 <$ P-value < 0.02 (0.015)

7.
H_0: $\sigma_1^2 = \sigma_2^2$
H_1: $\sigma_1^2 \neq \sigma_2^2$ (claim)

C. V. = 1.88 $\alpha = \frac{0.05}{2}$
d. f. N = 59   d. f. D = 59
$F = \frac{s_1^2}{s_2^2} = \frac{7.6^2}{5.4^2} = 1.981$

0   1.88   ↑
                                  1.981

Reject the null hypothesis. There is enough evidence to support the claim that there is a difference in the variances.

9.
H_0: $\sigma_1^2 = \sigma_2^2$
H_1: $\sigma_1^2 \neq \sigma_2^2$ (claim)

$s_1 = 2.2899$ $s_2 = 1.5859$
C. V. = 3.43 $\alpha = \frac{0.05}{2}$
d. f. N = 12   d. f. D = 11
$F = \frac{s_1^2}{s_2^2} = \frac{2.2899^2}{1.5859^2} = 2.085$

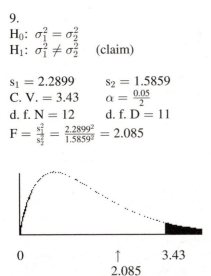

0   ↑   3.43
                2.085

Do not reject the null hypothesis. There is not enough evidence to support the claim that the variances are different.

11.
H_0: $\sigma_1^2 = \sigma_2^2$
H_1: $\sigma_1^2 \neq \sigma_2^2$ (claim)

$s_1 = 33.99$ $s_2 = 33.99$
C. V. = 4.99 $\alpha = \frac{0.05}{2}$
d. f. N = 7   d. f. D = 7
$F = \frac{s_1^2}{s_2^2} = \frac{(33.99)^2}{(33.99)^2} = 1$

11. continued

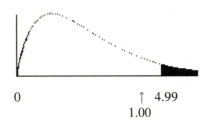

0   ↑ 4.99
                          1.00

Do not reject the null hypothesis. There is not enough evidence to support the claim that the variance in the number of calories differs between the two brands.

13.
H_0: $\sigma_1^2 = \sigma_2^2$
H_1: $\sigma_1^2 > \sigma_2^2$ (claim)

$s_1 = 111.211$ $s_2 = 35.523$
$n_1 = 7$ $n_2 = 6$
d. f. N = 6   d. f. D = 5
C. V. = 4.950 at $\alpha = 0.05$
C. V. = 10.67 at $\alpha = 0.01$

$F = \frac{s_1^2}{s_2^2} = \frac{(111.211)^2}{(35.523)^2} = 9.801$

0   4.95   ↑
                         9.801

Reject the null hypothesis at $\alpha = 0.05$. There is enough evidence to support the claim that the variance in area is greater for Eastern cities.

                  ↑ 10.67
                9.801

Do not reject the null hypothesis at $\alpha = 0.01$. There is not enough evidence to support the claim that the variance in area is greater for Eastern cities.

Chapter 9 - Testing the Difference Between
Two Means, Two Proportions, and Two Variances

15.
H_0: $\sigma_1^2 = \sigma_2^2$
H_1: $\sigma_1^2 \neq \sigma_2^2$ (claim)

Research: $s_1 = 5501.118$
Primary Care: $s_2 = 5238.809$

C. V. = 4.03 $\alpha = \frac{0.05}{2}$
d. f. N = 9   d. f. D = 9
$F = \frac{s_1^2}{s_2^2} = \frac{(5501.118)^2}{(5238.809)^2} = 1.10$

0   ↑   4.03
                    1.10

Do not reject the null hypothesis. There is
not enough evidence to support the claim
that there is a difference between the
variances in tuition costs.

17.
H_0: $\sigma_1^2 = \sigma_2^2$ (claim)
H_1: $\sigma_1^2 \neq \sigma_2^2$

$s_1 = 130.496$ $s_2 = 73.215$
C. V. = 3.87 $\alpha = \frac{0.10}{2}$
d. f. N = 6   d. f. D = 7
$F = \frac{s_1^2}{s_2^2} = \frac{(130.496)^2}{(73.215)^2} = 3.18$

0   ↑   3.87
                    3.18

Do not reject the null hypothesis. There is
not enough evidence to reject the claim that
the variances of the heights are equal.

19.

Men	Women
$s_1^2 = 2.363$	$s_2^2 = 0.444$
$n_1 = 15$	$n_2 = 15$

H_0: $\sigma_1^2 = \sigma_2^2$ (claim)
H_1: $\sigma_1^2 \neq \sigma_2$

19. continued
$\alpha = 0.05$ P-value = 0.004
d. f. N = 14   d. f. D = 14
$F = \frac{s_1^2}{s_2^2} = \frac{2.363}{0.444} = 5.32$

Since P-value $< \alpha$, reject the null
hypothesis. There is enough evidence to
reject the claim that the variances in weights
are equal.

REVIEW EXERCISES - CHAPTER 9

1.
H_0: $\mu_1 = \mu_2$
H_1: $\mu_1 > \mu_2$ (claim)

CV = 2.33 $\alpha = 0.01$
$\overline{X}_1 = 120.1$ $\overline{X}_2 = 117.8$
$s_1 = 16.722$ $s_2 = 16.053$

$z = \frac{(\overline{X}_1 - \overline{X}_2) - (\mu_1 - \mu_2)}{\sqrt{\frac{\sigma_1^2}{n_1} + \frac{\sigma_2^2}{n_2}}} = \frac{(120.1 - 117.8) - 0}{\sqrt{\frac{16.722^2}{36} + \frac{16.053^2}{35}}}$

$z = 0.587$

0   ↑   2.33
   0.59

Do not reject the null hypothesis. There is
not enough evidence to support the claim
that single people do more pleasure driving
than married people.

3.
H_0: $\mu_1 = \mu_2$
H_1: $\mu_1 > \mu_2$ (claim)

C. V. = 1.729 $\alpha = 0.05$
$t = \frac{(\overline{X}_1 - \overline{X}_2) - (\mu_1 - \mu_2)}{\sqrt{\frac{s_1^2}{n_1} + \frac{s_2^2}{n_2}}} = \frac{(16.7 - 12.5) - 0}{\sqrt{\frac{8.41}{26} + \frac{10.24}{20}}}$

$t = 4.595$

3. continued

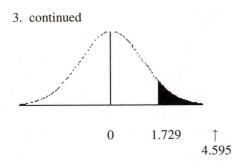

0 1.729 ↑
4.595

Reject the null hypothesis. There is enough evidence to support the claim that single persons spend a greater time each day communicating.

5.
H_0: $\mu_1 = \mu_2$
H_1: $\mu_1 \neq \mu_2$ (claim)

C. V. = ± 2.624 d. f. = 14

$t = \dfrac{(\overline{X}_1 - \overline{X}_2) - (\mu_1 - \mu_2)}{\sqrt{\frac{s_1^2}{n_1} + \frac{s_2^2}{n_2}}} = \dfrac{(35{,}270 - 29{,}512) - 0}{\sqrt{\frac{3256^2}{15} + \frac{1432^2}{15}}}$

$t = 6.54$

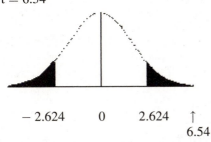

-2.624 0 2.624 ↑
6.54

Reject the null hypothesis. There is enough evidence to support the claim that there is a difference in the teachers' salaries.

98% Confidence Interval:
$\$3{,}494.80 < \mu_1 - \mu_2 < \$8{,}021.20$

7.

Maximum	Minimum	D	D^2
44	27	17	289
46	34	12	144
46	24	22	484
36	19	17	289
34	19	15	225
36	26	10	100
57	33	24	576
62	57	5	25
73	46	27	729
53	26	27	729
		$\sum D = 176$	$\sum D^2 = 3590$

7. continued
H_0: $\mu_D = 10$
H_1: $\mu_D > 10$

C. V. = 2.821
$\overline{D} = \frac{176}{10} = 17.6$

$s_D = \sqrt{\dfrac{3590 - \frac{176^2}{10}}{9}} = 7.3967$

$t = \dfrac{17.6 - 10}{\frac{7.3967}{\sqrt{10}}} = 3.249$

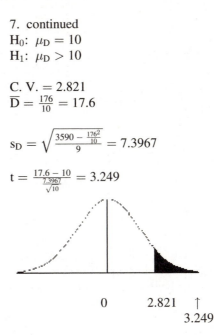

0 2.821 ↑
3.249

Reject the null hypothesis. There is enough evidence to support the claim that the mean difference is more than 10 degrees.

9.
$\hat{p}_1 = \frac{49}{200} = 0.245$ $\hat{p}_2 = \frac{62}{200} = 0.31$

$\overline{p} = \frac{49 + 62}{200 + 200} = 0.2775$
$\overline{q} = 1 - 0.2775 = 0.7225$

H_0: $p_1 = p_2$
H_1: $p_1 \neq p_2$ (claim)

C. V. = ± 1.96
$z = \dfrac{(0.245 - 0.31) - 0}{\sqrt{(0.2775)(0.7225)(\frac{1}{200} + \frac{1}{200})}} = -1.45$

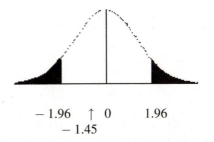

-1.96 ↑ 0 1.96
-1.45

Do not reject the null hypothesis. There is not enough evidence to support the claim that the proportions are different.

11.
H_0: $\sigma_1 = \sigma_2$
H_1: $\sigma_1 \neq \sigma_2$ (claim)

11. continued

$\alpha = 0.10$

d.f.N. $= 23$ d.f.D. $= 10$

C. V. $= 2.77$

$F = \frac{13.2^2}{4.1^2} = 10.365$

2.77   ↑

10.365

Reject the null hypothesis. There is enough evidence to support the claim that there is a difference in the standard deviations.

13.

H_0: $\sigma_1^2 = \sigma_2^2$

H_1: $\sigma_1^2 \neq \sigma_2^2$ (claim)

C. V. $= 2.45$ $\alpha = \frac{0.05}{2}$

d.f.N. $= 8$ d.f.D. $= 7$

$F = \frac{5.21^2}{4.08^2} = 1.631$

0   ↑ 2.45

1.631

Do not reject the null hypothesis. There is not enough evidence to support the claim that there is a difference in the variances.

To conclude a difference:

$\frac{5.21^2}{x^2} > 2.45$

$27.1441 > 2.45x^2$

$11.08 > x^2$

$\sqrt{11.08} > \sqrt{x^2}$

$\$3.33 > x$

Thus, store Z's standard deviation would have to be less than \$3.33 in order to conclude a difference in the variances.

CHAPTER 9 QUIZ

1. False, there are different formulas for independent and dependent samples.
2. False, the samples are independent.
3. True
4. False, they can be right, left, or two tailed.
5. d
6. a
7. c
8. a
9. $\mu_1 = \mu_2$
10. t
11. Normal
12. Negative
13. $\frac{s_1^2}{s_2^2}$

14. H_0: $\mu_1 = \mu_2$

H_1: $\mu_1 \neq \mu_2$ (claim)

C. V. $= \pm 2.58$ $z = -3.69$

Reject the null hypothesis. There is enough evidence to support the claim that there is a difference in the cholesterol levels of the two groups.

99% Confidence Interval:

$-10.2 < \mu_1 - \mu_2 < -1.8$

15. H_0: $\mu_1 = \mu_2$

H_1: $\mu_1 > \mu_2$ (claim)

C. V. $= 1.28$ $z = 1.60$

Reject the null hypothesis. There is enough evidence to support the claim that average rental fees for the east apartments is greater than the average rental fees for the west apartments.

16. H_0: $\mu_1 = \mu_2$

H_1: $\mu_1 \neq \mu_2$ (claim)

$t = 11.094$ C. V. $= \pm 2.779$

Reject the null hypothesis. There is enough evidence to support the claim that the average prices are different.

99% Confidence Interval:

$0.298 < \mu_1 - \mu_2 < 0.502$

(TI: $0.2995 < \mu_1 - \mu_2 < 0.5005$)

17. H_0: $\mu_1 = \mu_2$

H_1: $\mu_1 < \mu_2$ (claim)

C. V. $= -1.860$ d.f. $= 8$ t $= -4.05$

Reject the null hypothesis. There is enough evidence to support the claim that accidents have increased.

18. H_0: $\mu_1 = \mu_2$

H_1 $\mu_1 \neq \mu_2$ (claim)

18. continued
C. V. $= \pm 2.718$ t $= 9.807$
Reject the null hypothesis. There is enough evidence to support the claim that the salaries are different.

98% Confidence Interval:
$\$6653 < \mu_1 - \mu_2 < \$11,757$
(TI: $\$6619 < \mu_1 - \mu_2 < \$11,491$)

19. H_0: $\mu_1 = \mu_2$
H_1 $\mu_1 > \mu_2$ (claim)
d. f. $= 10$ $\qquad$ t $= 0.874$
$0.10 <$ P-value < 0.25 (0.198)
Do not reject the null hypothesis since P-value > 0.05. There is not enough evidence to support the claim that the incomes of city residents are greater than the incomes of rural residents.

20. H_0: $\mu_D = 0$
H_1 $\mu_D < 0$ (claim)
C. V. $= -2.821$ t $= -4.17$
Do not reject the null hypothesis. There is not enough evidence to support the claim that the sessions improved math skills.

21. H_0: $\mu_D = 0$
H_1 $\mu_D < 0$ (claim)
C. V. $= -1.833$ t $= -1.71$
Reject the null hypothesis. There is enough evidence to support the claim that egg production was increased.

22. H_0: $p_1 = p_2$
H_1: $p_1 \neq p_2$ (claim)
C. V. $= \pm 1.65$ z $= -0.69$
Do not reject the null hypothesis. There is not enough evidence to support the claim that the proportions are different.

90% Confidence Interval:
$-0.105 < p_1 - p_2 < 0.045$

23. H_0: $p_1 = p_2$ (claim)
H_1: $p_1 \neq p_2$
C. V. $= \pm 1.96$ z $= 0.544$
Do not reject the null hypothesis. There is not enough evidence to support the claim that the proportions have changed.

95% Confidence Interval:
$-0.026 < p_1 - p_2 < 0.0460$

23. continued
Yes, the confidence interval contains 0; hence, the null hypothesis is not rejected.

24. H_0: $\sigma_1^2 = \sigma_2^2$
H_1 $\sigma_1^2 \neq \sigma_2^2$ (claim)
d. f. N. $= 17$ $\qquad$ d. f. D. $= 14$
$F = 1.637$
P-value > 0.20 (0.357)
Do not reject since P-value > 0.05. There is not enough evidence to support the claim that the variances are different.

25. H_0: $\sigma_1^2 = \sigma_2^2$
H_1 $\sigma_1^2 \neq \sigma_2^2$
C. V. $= 1.90$ $\qquad$ $F = 1.296$
Do not reject. There is not enough evidence to support the claim that the variances are different.

Note: Graphs are not to scale and are intended to convey a general idea.

Answers may vary due to rounding, TI-83's, or computer programs.

EXERCISE SET 10-1

1.
Two variables are related when there exists a discernible pattern between them.

3.
r, ρ (rho)

5.
A positive relationship means that as x increases, y also increases.
A negative relationship means that as x increases, y decreases.

7.
Answers will vary.

9.
Pearson's Product Moment Correlation Coefficient.

11.
There are many other possibilities, such as chance, relationship to a third variable, etc.

13.

$\sum x = 1045$
$\sum y = 9283$
$\sum x^2 = 299{,}315$
$\sum y^2 = 21{,}881{,}839$
$\sum xy = 2{,}380{,}435$
$n = 9$

$$r = \frac{n(\sum xy)-(\sum x)(\sum y)}{\sqrt{[n(\sum x^2)-(\sum x)^2]\,[n(\sum y^2)-(\sum y)^2]}}$$

$$r = \frac{9(2{,}380{,}435)-(1045)(9283)}{\sqrt{[9(299{,}315)-(1045)^2]\,[9(21{,}881{,}839)-(9283)^2]}}$$

13. continued
$r = 0.880$

$H_0: \rho = 0$
$H_1: \rho \neq 0$
C. V. $= \pm 0.666$ d. f. $= 7$

Decision: Reject. There is a significant linear relationship between number of movie releases and gross receipts.

15.

$\sum x = 32$
$\sum y = 1105$
$\sum x^2 = 220$
$\sum y^2 = 364{,}525$
$\sum xy = 3405$
$n = 6$

$$r = \frac{n(\sum xy)-(\sum x)(\sum y)}{\sqrt{[n(\sum x^2)-(\sum x)^2]\,[n(\sum y^2)-(\sum y)^2]}}$$

$$r = \frac{6(3405)-(32)(1105)}{\sqrt{[6(220)-(32)^2][6(364525)-(1105)^2]}}$$

$r = -0.883$

$H_0: \rho = 0$
$H_1: \rho \neq 0$
C. V. $= \pm 0.811$ d. f. $= 4$
Decision: Reject. There is a significant linear relationship between the number of years a person has been out of school and his or her contributions.

17.

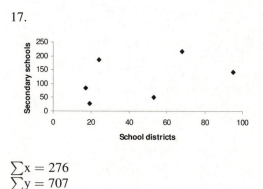

$\sum x = 276$
$\sum y = 707$

104

17. continued
$$\sum x^2 = 17{,}684$$
$$\sum y^2 = 112{,}359$$
$$\sum xy = 37{,}352$$
n = 6
$$r = \frac{n(\sum xy)-(\sum x)(\sum y)}{\sqrt{[n(\sum x^2)-(\sum x)^2][n(\sum y^2)-(\sum y)^2]}}$$

$$r = \frac{6(37{,}352)-(276)(707)}{\sqrt{[6(17{,}684)-276^2][6(112{,}359)-707^2]}}$$

$$r = 0.401$$

H_0: $\rho = 0$
H_1: $\rho \neq 0$
C. V. = ± 0.811 d. f. = 4
Decision: Do not reject. There is no significant linear relationship between the number of school districts and the number of secondary schools.

19.

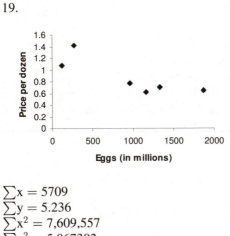

$$\sum x = 5709$$
$$\sum y = 5.236$$
$$\sum x^2 = 7{,}609{,}557$$
$$\sum y^2 = 5.067302$$
$$\sum xy = 4115.025$$
n = 6
$$r = \frac{6(4115.025)-(5709)(5.236)}{\sqrt{[6(7{,}609{,}557-(5709)^2][6(5.067302)-(5.236)^2]}}$$

$$r = -0.833$$

H_0: $\rho = 0$
H_1: $\rho \neq 0$
C. V. = ± 0.811 d. f. = 4
Decision: Reject. There is a significant linear relationship between the number of eggs produced and price per dozen.

21.

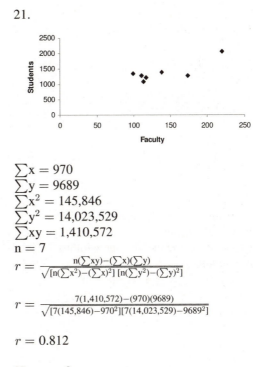

$$\sum x = 970$$
$$\sum y = 9689$$
$$\sum x^2 = 145{,}846$$
$$\sum y^2 = 14{,}023{,}529$$
$$\sum xy = 1{,}410{,}572$$
n = 7
$$r = \frac{n(\sum xy)-(\sum x)(\sum y)}{\sqrt{[n(\sum x^2)-(\sum x)^2][n(\sum y^2)-(\sum y)^2]}}$$

$$r = \frac{7(1{,}410{,}572)-(970)(9689)}{\sqrt{[7(145{,}846)-970^2][7(14{,}023{,}529)-9689^2]}}$$

$$r = 0.812$$

H_0: $\rho = 0$
H_1: $\rho \neq 0$
C. V. = ± 0.754 d. f. = 5
Decision: Reject. There is a significant linear relationship between the number of faculty and the number of students.

For x = Students and y = Faculty:

$$r = \frac{7(1{,}410{,}572)-(9689)(970)}{\sqrt{[7(14{,}023{,}529)-9689^2][7(145{,}846)-970^2]}}$$

$$r = 0.812$$

The results are the same when x and y are switched. Students would be the more likely independent variable.

23.

$$\sum x = 557$$
$$\sum y = 17.7$$
$$\sum x^2 = 44{,}739$$
$$\sum y^2 = 55.99$$

23. continued

$\sum xy = 1468.9$

$n = 7$

$r = \dfrac{n(\sum xy)-(\sum x)(\sum y)}{\sqrt{[n(\sum x^2)-(\sum x)^2]\,[n(\sum y^2)-(\sum y)^2]}}$

$r = \dfrac{7(1468.9) - (557)(17.7)}{\sqrt{[7(44,739) - 557^2][7(55.99) - 17.7^2]}}$

$r = 0.883$

H_0: $\rho = 0$
H_1: $\rho \neq 0$
C. V. $= \pm 0.754$ d. f. $= 5$
Decision: Reject. There is a significant linear relationship between the average daily temperature and the average monthly precipitation.

25.

$\sum x = 2040$
$\sum y = 97$
$\sum x^2 = 790,200$
$\sum y^2 = 1861$
$\sum xy = 38,260$
$n = 6$

$r = \dfrac{n(\sum xy)-(\sum x)(\sum y)}{\sqrt{[n(\sum x^2)-(\sum x)^2]\,[n(\sum y^2)-(\sum y)^2]}}$

$r = \dfrac{6(38,260)-(2040)(97)}{\sqrt{[6(790,200)-(2040)^2][6(1861)-(97)^2]}}$

$r = 0.993$

H_0: $\rho = 0$
H_1: $\rho \neq 0$
C. V. $= \pm 0.811$ d. f. $= 4$
Decision: Reject. There is a significant linear relationship between the number of fat calories and the fat grams.

27.

$\sum x = 1013$
$\sum y = 748$
$\sum x^2 = 168,435$
$\sum y^2 = 90,626$
$\sum xy = 120,953$
$n = 7$

$r = \dfrac{7(120,953)-(1013)(748)}{\sqrt{[7(168,435)-(1013)^2]\,[7(90,626)-(748)^2]}}$

$r = 0.831$

H_0: $\rho = 0$
H_1: $\rho \neq 0$
C. V. $= \pm 0.754$ d. f. $= 5$
Decision: Reject. There is a significant linear relationship between the number of licensed beds in a hospital and the number of staffed beds.

29.

$r = \dfrac{n(\sum xy)-(\sum x)(\sum y)}{\sqrt{[n(\sum x^2)-(\sum x)^2]\,[n(\sum y^2)-(\sum y)^2]}}$

$r = \dfrac{5(125)-(15)(35)}{\sqrt{[5(55)-(15)^2][5(285)-(35)^2]}} = 1$

$r = \dfrac{5(125)-(35)(15)}{\sqrt{[5(285)-(35)^2][5(55)-(15)^2]}} = 1$

The value of r does not change when the values for x and y are interchanged.

EXERCISE 10-2

1.
Draw the scatter plot and test the significance of the correlation coefficient.

3.
$y' = a + bx$

5.
It is the line that is drawn through the points on the scatter plot such that the sum of the squares of the vertical distances each point is from the line is at a minimum.

7.
When r is positive, b will be positive. When r is negative, b will be negative.

9.
The closer r is to $+1$ or -1, the more accurate the predicted value will be.

11.
When r is not significant, the mean of the y values should be used to predict y.

13.
$$a = \frac{(\sum y)(\sum x^2) - (\sum x)(\sum xy)}{n(\sum x^2) - (\sum x)^2}$$

$$a = \frac{(9283)(299,315) - (1045)(2,380,435)}{9(299,315) - (1045)^2} = 181.661$$

$$b = \frac{n(\sum xy) - (\sum x)(\sum y)}{n(\sum x^2) - (\sum x)^2}$$

$$b = \frac{9(2,380,435) - (1045)(9283)}{9(299,315) - (1045)^2} = 7.319$$

$y' = a + bx$
$y' = 181.661 + 7.319x$
$y' = 181.661 + 7.319(200) = \$1645.5$
million gross receipts

15.
$$a = \frac{(\sum y)(\sum x^2) - (\sum x)(\sum xy)}{n(\sum x^2) - (\sum x)^2}$$

$$a = \frac{(1105)(220) - (32)(3405)}{6(220) - (32)^2}$$

$$a = \frac{243100 - 108960}{1320 - 1024} = \frac{134140}{296} = 453.176$$

$$b = \frac{n(\sum xy) - (\sum x)(\sum y)}{n(\sum x^2) - (\sum x)^2}$$

$$b = \frac{6(3405) - (32)(1105)}{6(220) - (32)^2} = \frac{20430 - 35360}{296}$$

$$b = \frac{-14930}{296} = -50.439$$

$y' = a + bx$
$y' = 453.176 - 50.439x$
$y' = 453.176 - 50.439(4) = \$251.42$

17.
Since r is not significant, no regression should be done.

19.
$$a = \frac{(\sum y)(\sum x^2) - (\sum x)(\sum xy)}{n(\sum x^2) - (\sum x)^2}$$

$$a = \frac{(5.236)(7,609,557) - (5709)(4115.025)}{6(7,609,557) - (5709)^2}$$

19. continued
$a = 1.252$

$$b = \frac{n(\sum xy) - (\sum x)(\sum y)}{n(\sum x^2) - (\sum x)^2}$$

$$b = \frac{6(4115.025) - (5709)(5.236)}{6(4115.025) - (5709)^2}$$

$$b = -0.000398$$

$y' = a + bx$
$y' = 1.252 - 0.000398x$
$y' = 1.252 - 0.000398(1600) = 0.615$
per dozen

21.
For $x =$ Students and $y =$ Faculty:

$$a = \frac{(\sum y)(\sum x^2) - (\sum x)(\sum xy)}{n(\sum x^2) - (\sum x)^2}$$

$$a = \frac{(970)(14,023,529) - (9689)(1,410,572)}{7(14,023,529) - (9689)^2}$$

$a = -14.974$

$$b = \frac{n(\sum xy) - (\sum x)(\sum y)}{n(\sum x^2) - (\sum x)^2}$$

$$b = \frac{7(1,410,572) - (9689)(970)}{7(14,023,529) - (9689)^2} = 0.111$$

$y' = a + bx$
$y' = -14.974 + 0.111x$

23.
$$a = \frac{(\sum y)(\sum x^2) - (\sum x)(\sum xy)}{n(\sum x^2) - (\sum x)^2}$$

$$a = \frac{(17.7)(44739) - (557)(1468.9)}{7(44739) - (557)^2} = -8.994$$

$$b = \frac{n(\sum xy) - (\sum x)(\sum y)}{n(\sum x^2) - (\sum x)^2}$$

$$b = \frac{7(1468.9) - (557)(17.7)}{7(44739) - (557)^2} = 0.1448$$

$y' = a + bx$
$y' = -8.994 + 0.1448x$
$y' = -8.994 + 0.1448(70) = 1.1$ inches

25.
$$a = \frac{(\sum y)(\sum x^2) - (\sum x)(\sum xy)}{n(\sum x^2) - (\sum x)^2}$$

$$a = \frac{(97)(790,200) - (2040)(38,260)}{6(790,200) - (2040)^2}$$

$a = -2.417$

25. continued

$$b = \frac{n(\sum xy) - (\sum x)(\sum y)}{n(\sum x^2) - (\sum x)^2}$$

$$b = \frac{6(38,260) - (2040)(97)}{6(790,200) - (2040)^2} = 0.055$$

$y\prime = a + bx$
$y\prime = -2.417 + 0.055x$
$y\prime = -2.417 + 0.055(400) = 19.6$ grams

27.

$$a = \frac{(\sum y)(\sum x^2) - (\sum x)(\sum xy)}{n(\sum x^2) - (\sum x)^2}$$

$$a = \frac{(748)(168,435) - (1013)(120,953)}{7(168,435) - (1013)^2}$$
$a = 22.659$

$$b = \frac{n(\sum xy) - (\sum x)(\sum y)}{n(\sum x^2) - (\sum x)^2}$$

$$b = \frac{7(120,953) - (1013)(748)}{7(168,435) - (1013)^2} = 0.582$$

$y' = a + bx$
$y' = 22.659 + 0.582x$
$y' = 22.659 + 0.582(44) = 48.267$ staffed beds

29.

$\sum x = 285$
$\sum y = 1289$
$\sum x^2 = 15,637.88$
$\sum y^2 = 305,731$
$\sum xy = 64,565.8$
$n = 6$

$$r = \frac{n(\sum xy) - (\sum x)(\sum y)}{\sqrt{[n(\sum x^2) - (\sum x)^2][n(\sum y^2) - (\sum y)^2]}}$$

$$r = \frac{6(64,565.8) - (285)(1289)}{\sqrt{[6(15,637.88) - (285)^2][6(305,731) - (1289)^2]}}$$

$r = 0.429$

$H_0: \rho = 0$
$H_1: \rho \neq 0$
C. V. $= \pm 0.811$ d. f. $= 4$

29. continued
Decision: Do not reject
There is no significant relationship between the number of farms and acreage.

31.

Tons of Coal vs. Number of Employees

$y' = 34.852 + 0.140x$

$\sum x = 4027$
$\sum y = 26,728$
$\sum x^2 = 3,550,103$
$\sum y^2 = 162,101,162$
$\sum xy = 23,663,669$
$n = 8$

$$r = \frac{n(\sum xy) - (\sum x)(\sum y)}{\sqrt{[n(\sum x^2) - (\sum x)^2][n(\sum y^2) - (\sum y)^2]}}$$

$$r = \frac{8(23662669) - (4027)(26728)}{\sqrt{[8(3550103) - 4027^2][8(162101162) - (26728)^2]}}$$

$r = 0.970$

$H_0: \rho = 0$
$H_1: \rho \neq 0$
C. V. $= \pm 0.707$ d. f. $= 6$
Decision: Reject. There is a significant relationship between the number of tons of coal produced and the number of employees.

$$a = \frac{(\sum y)(\sum x^2) - (\sum x)(\sum xy)}{n(\sum x^2) - (\sum x)^2}$$

$$a = \frac{(26728)(3550103) - (4027)(23663669)}{8(3550103) - (4027)^2}$$

$a = -33.358$

$$b = \frac{n(\sum xy) - (\sum x)(\sum y)}{n(\sum x^2) - (\sum x)^2}$$

$$b = \frac{8(23663669) - (4027)(26728)}{8(3550103) - (4027)^2} = 6.703$$

$y' = a + bx$
$y' = -33.358 + 6.703x$
$y' = -33.358 + 6.703(500) = 3318.142$

33.

$\sum x = 37$
$\sum y = 482$
$\sum x^2 = 337$
$\sum y^2 = 39526$
$\sum xy = 2682$
n = 6

$$r = \frac{n(\sum xy)-(\sum x)(\sum y)}{\sqrt{[n(\sum x^2)-(\sum x)^2][n(\sum y^2)-(\sum y)^2]}}$$

$$r = \frac{6(2682)-(37)(482)}{\sqrt{[6(337)-(37)^2][6(39526)-(482)^2]}}$$

$r = -0.981$

H_0: $\rho = 0$
H_1: $\rho \neq 0$
C. V. = ± 0.811 d. f. = 4
Decision: Reject. There is a significant relationship between the number of absences and the final grade.

$$a = \frac{(\sum y)(\sum x^2)-(\sum x)(\sum xy)}{n(\sum x^2)-(\sum x)^2}$$

$$a = \frac{(482)(337)-(37)(2682)}{6(337)-(37)^2} = 96.784$$

$$b = \frac{n(\sum xy)-(\sum x)(\sum y)}{n(\sum x^2)-(\sum x)^2}$$

$$b = \frac{6(2682)-(37)(482)}{6(337)-(37)^2} = -2.668$$

$y' = a + bx$
$y' = 96.784 - 2.668x$

35.

$\sum x = 580$
$\sum y = 108$
$\sum x^2 = 35,780$
$\sum y^2 = 1304$
$\sum xy = 6120$
n = 10

$$r = \frac{n(\sum xy)-(\sum x)(\sum y)}{\sqrt{[n(\sum x^2)-(\sum x)^2][n(\sum y^2)-(\sum y)^2]}}$$

$$r = \frac{10(6120)-(580)(108)}{\sqrt{10(35,780)-580^2][10(1304)-108^2]}}$$

$r = -0.265$
H_0: $\rho = 0$
H_1: $\rho \neq 0$
P-value > 0.05 (0.459)
Decision: Do not reject since P-value > 0.05. There is no significant linear relationship between the ages of billionaires and their net worth. Since r is not significant, no regression should be done.

37.
For Exercise 15:
$\bar{x} = 5.3333$
$\bar{y} = 184.1667$
$b = -50.439$
$a = \bar{y} - b\bar{x}$
$a = 184.1667 - (-50.439)(5.3333)$
$a = 184.1667 + 269.0063$
$a = 453.173$ (differs due to rounding)

For Exercise 16:
Since r is not significant, no regression should be done.

EXERCISE SET 10-3

1.
Explained variation is the variation due to the relationship and is computed by $\sum(y' - \bar{y})^2$.

3.
Total variation is the sum of the squares of the vertical distances of the points from the mean. It is computed by $\sum(y - \bar{y})^2$.

5.
It is found by squaring r.

7.
The coefficient of non-determination is $1 - r^2$.

9.
For $r = 0.75$, $r^2 = 0.5625$ and $1 - r^2 = 0.4375$.
Thus 56.25% of the variation of y is due to the variation of x, and 43.75% of the variation of y is due to chance.

11.
For $r = 0.42$, $r^2 = 0.1764$ and $1 - r^2 = 0.8236$. Thus 17.64% of the variation of y is due to the variation of x, and 82.36% of the variation of y is due to chance.

13.
For $r = 0.91$, $r^2 = 0.8281$ and $1 - r^2 = 0.1719$. Thus 82.81% of the variation of y is due to the variation of x, and 17.19% of the variation of y is due to chance.

Note: For Exercises 15 − 17, values for a and b are rounded to 3 decimal places according to the textbook's rounding rule for intercept and slope of the regression equation. Where these answers differ from the text, additional decimal places are included to show consistency with text answers.

15.
$$S_{est} = \sqrt{\frac{\sum y^2 - a\sum y - b\sum xy}{n-2}}$$

$$S_{est} = \sqrt{\frac{21,881,839 - 181.661(9283) - (7.319)(2,380,435)}{9-2}}$$

$$S_{est} = \sqrt{396,153.7389} = 629.41$$

Using $a = 181.661102$ and $b = 7.318708$,
$S_{est} = 629.4862$

17.
$$S_{est} = \sqrt{\frac{\sum y^2 - a\sum y - b\sum xy}{n-2}} =$$
$$S_{est} = \sqrt{\frac{364525 - (453.176)(1105) - (-50.439)(3405)}{6-2}}$$
$$S_{est} = 94.22$$

19.
$y' = 181.661 + 7.319x$
$y' = 181.661 + 7.319(200)$
$y' = 1645.461$

$$y' - t_{\frac{\alpha}{2}} \cdot S_{est}\sqrt{1 + \frac{1}{n} + \frac{n(x-\bar{X})}{n\sum x^2 - (\sum x)^2}} < y <$$

$$y' + t_{\frac{\alpha}{2}} \cdot S_{est}\sqrt{1 + \frac{1}{n} + \frac{n(x-\bar{X})^2}{n\sum x^2 - (\sum x)^2}}$$

$$1645.461 - (1.895)(629.4862)\sqrt{1 + \frac{1}{9} + \frac{9(200-116.11)^2}{9(299,315)-1045^2}}$$
$$< y < 1645.461 +$$
$$(1.895)(629.4862)\sqrt{1 + \frac{1}{9} + \frac{9(200-116.11)^2}{9(299,315)-1045^2}}$$

$$1645.461 - 1279.580227 < y <$$
$$1645.461 + 1279.580227$$

$$365.88 < y < 2925.04$$

21.
$y' = 453.176 - 50.439x$
$y' = 453.176 - 50.439(4)$
$y' = 251.42$
$$y' - t_{\frac{\alpha}{2}} \cdot S_{est}\sqrt{1 + \frac{1}{n} + \frac{n(x-\bar{X})^2}{n\sum x^2 - (\sum x)^2}} < y <$$

$$y' + t_{\frac{\alpha}{2}} \cdot S_{est}\sqrt{1 + \frac{1}{n} + \frac{n(x-\bar{X})^2}{n\sum x^2 - (\sum x)^2}}$$

$$251.42 - 2.132(94.22)\sqrt{1 + \frac{1}{6} + \frac{6(4-5.33)^2}{6(220)-32^2}}$$

$$< y < 251.42 + 2.132(94.22)\sqrt{1 + \frac{1}{6} + \frac{6(4-5.33)^2}{6(220)-32^2}}$$

$$251.42 - (2.132)(94.22)(1.1) < y <$$
$$251.42 + (2.132(94.22)(1.1)$$
$$\$30.46 < y < \$472.38$$

EXERCISE SET 10-4

1.
Simple linear regression has one independent variable and one dependent variable. Multiple regression has one dependent variable and two or more independent variables.

3.
The relationship would include all variables.

5.
The multiple correlation coefficient R is always higher than the individual correlation coefficients. Also, the value of R can range from 0 to $+1$.

7.
$y' = 9.6 + 2.2x_1 - 1.08x_2$
$y' = 9.6 + 2.2(9) - 1.08(24) = 3.48$

9.
$y' = -14.9 + 0.93359x_1 + 0.99847x_2$
$\qquad + 5.3844x_3$
$y' = -14.9 + 0.93359(8) + 0.99847(34)$
$\qquad + 5.3844(11)$
$y' = 85.75$ (grade) or 86

11.
R is a measure of the strength of the relationship between the dependent variables and all the independent variables.

13.
R^2 is the coefficient of multiple determination. R^2_{adj} is adjusted for sample size and the number of predictors.

15.
The F test is used to test the significance of R.

REVIEW EXERCISES - CHAPTER 10

1.

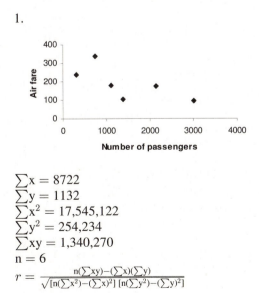

$\sum x = 8722$
$\sum y = 1132$
$\sum x^2 = 17,545,122$
$\sum y^2 = 254,234$
$\sum xy = 1,340,270$
$n = 6$
$r = \dfrac{n(\sum xy) - (\sum x)(\sum y)}{\sqrt{[n(\sum x^2) - (\sum x)^2]\,[n(\sum y^2) - (\sum y)^2]}}$

1. continued
$r = \dfrac{6(1,340,270) - (8722)(1132)}{\sqrt{[6(17,545,122) - (8722)^2][6(254,234) - (1132)^2]}}$

$r = -0.686$

H_0: $\rho = 0$
H_1: $\rho \neq 0$
C. V. $= \pm 0.917$ at $\alpha = 0.01$ d. f. $= 4$

Decision: Do not reject. There is no significant linear relationship between the number of passengers and airfare.

3.

$\sum x = 175$
$\sum y = 631$
$\sum x^2 = 4667$
$\sum y^2 = 57,335$
$\sum xy = 16,093$
$n = 7$
$r = \dfrac{n(\sum xy) - (\sum x)(\sum y)}{\sqrt{[n(\sum x^2) - (\sum x)^2]\,[n(\sum y^2) - (\sum y)^2]}}$

$r = \dfrac{7(16,093) - (175)(631)}{\sqrt{[7(4667) - (175)^2][7(57,335) - (631)^2]}}$

$r = 0.873$

H_0: $\rho = 0$
H_1: $\rho \neq 0$
C. V. $= \pm 0.875$ d. f. $= 5$

Decision: Do not reject. There is not a significant relationship between touchdowns and quarterback ratings.

5.

$\sum x = 884$
$\sum y = 47.8$
$\sum x^2 = 67,728$
$\sum y^2 = 242.06$
$\sum xy = 3163.8$
n = 12

$r = \dfrac{n(\sum xy) - (\sum x)(\sum y)}{\sqrt{[n(\sum x^2) - (\sum x)^2]\,[n(\sum y^2) - (\sum y)^2]}}$

$r = \dfrac{12(3163.8) - (884)(47.8)}{\sqrt{[12(67728) - (884)^2][12(242.06) - (47.8)^2]}}$

$r = -0.974$

H_0: $\rho = 0$
H_1: $\rho \neq 0$
C. V. $= \pm 0.708$ d. f. = 10

Decision: Reject. There is a significant relationship between speed and time.

$a = \dfrac{(\sum y)(\sum x^2) - (\sum x)(\sum xy)}{n(\sum x^2) - (\sum x)^2}$

$a = \dfrac{(47.8)(67728) - (884)(3163.8)}{12(67728) - (884)^2}$

$a = 14.086$

$b = \dfrac{n(\sum xy) - (\sum x)(\sum y)}{n(\sum x^2) - (\sum x)^2}$

$b = \dfrac{12(3163.8) - (884)(47.8)}{12(67728) - (884)^2}$

$b = -0.137$

$y' = a + bx$
$y' = 14.086 - 0.137x$
$y' = 14.086 - 0.137(72) = 4.222$

7.

$\sum x = 14,301$
$\sum y = 49,452$
$\sum x^2 = 49,158,943$
$\sum y^2 = 630,840,804$
$\sum xy = 168,984,784$
n = 7

$r = \dfrac{n(\sum xy) - (\sum x)(\sum y)}{\sqrt{[n(\sum x^2) - (\sum x)^2]\,[n(\sum y^2) - (\sum y)^2]}}$

$r = \dfrac{7(168,984,784) - (14,301)(49,452)}{\sqrt{[7(49,158,943) - (14,301)^2][7(630,840,804) - (49,452)^2]}}$

$r = 0.907$

H_0: $\rho = 0$
H_1: $\rho \neq 0$
C. V. ± 0.875 d. f. = 5

Decision: Reject. There is a significant relationship between female specialties and male specialties.

$a = \dfrac{(\sum y)(\sum x^2) - (\sum x)(\sum xy)}{n(\sum x^2) - (\sum x)^2}$

$a = \dfrac{(49,452)(49,158,943) - (14,301)(168,984,784)}{7(49,158,943) - (14,301)^2}$

$a = 102.846$

$b = \dfrac{n(\sum xy) - (\sum x)(\sum y)}{n(\sum x^2) - (\sum x)^2}$

$b = \dfrac{7(168,984,784) - (14,301)(49,452)}{7(49,158,943) - (14,301)^2}$

$b = 3.408$

$y' = a + bx$
$y' = 102.846 + 3.408x$
$y' = 102.846 + 3.408(2000)$
$y' = 6918.8$ or 6919

9.
$S_{est} = \sqrt{\dfrac{\sum y^2 - a\sum y - b\sum xy}{n - 2}}$

9. continued

$$S_{est} = \sqrt{\frac{242.06 - 14.086(47.8) + 0.137(3163.8)}{12-2}}$$

$$S_{est} = \sqrt{\frac{2.1898}{10}} = \sqrt{0.21898} = 0.468$$

(Note: TI-83 calculator answer is 0.513)

11.
(For calculation purposes only, since no regression should be done.)

$y' = 14.086 - 0.137x$
$y' = 14.086 - 0.137(72) = 4.222$

$$y' - t_{\frac{\alpha}{2}} \cdot S_{est} \sqrt{1 + \frac{1}{n} + \frac{n(x - \overline{X})^2}{n\Sigma x^2 - (\Sigma x)^2}} < y <$$

$$y' + t_{\frac{\alpha}{2}} \cdot S_{est} \sqrt{1 + \frac{1}{n} + \frac{n(x - \overline{X})^2}{n\Sigma x^2 - (\Sigma x)^2}}$$

$$4.222 - 1.812(0.468)\sqrt{1 + \frac{1}{12} + \frac{12(72 - 73.667)^2}{12(67,728) - 884^2}}$$
$$< y < 4.222 + 1.812(0.468)\sqrt{1 + \frac{1}{12} + \frac{12(72 - 73.667)^2}{12(67,728) - 884^2}}$$

$4.222 - 1.812(0.468)(1.041) < y <$
$\qquad 4.222 + 1.812(0.468)(1.041)$
$3.34 < y < 5.10$

13.
$y' = 12.8 + 2.09X_1 + 0.423X_2$
$y' = 12.8 + 2.09(4) + 0.423(2) = 22.006$ or
22.01

15.
$$R_{adj}^2 = 1 - \left[\frac{(1 - R^2)(n - 1)}{n - k - 1}\right]$$

$$R_{adj}^2 = 1 - \left[\frac{(1 - 0.873^2)(10 - 1)}{10 - 3 - 1}\right]$$

$$R_{adj}^2 = 1 - \left[\frac{2.1408}{6}\right] = 0.643$$

CHAPTER 10 QUIZ

1. False, the y variable would decrease.
2. True
3. True
4. False, the relationship may be affected by another variable, or by chance.
5. False, a relationship may be caused by chance.
6. False, there are several independent variables and one dependent variable.
7. a

8. a
9. d
10. c
11. b
12. Scatter plot
13. Independent
14. $-1, +1$
15. b (slope)
16. Line of best fit
17. $+1, -1$

18.

$\sum x = 18.61$
$\sum x^2 = 51.1919$
$\sum y = 7.67$
$\sum y^2 = 9.2083$
$\sum xy = 21.0956$
$n = 7$
$r = 0.600$
$H_0: \rho = 0$
$H_1: \rho \neq 0$
C.V. $= \pm 0.754$ d. f. $= 5$
Do not reject. There is no significant linear relationship between the price of the same drugs in the United States and in Australia. No regression should be done.

19.

$\sum x = 442$
$\sum x^2 = 27,964$
$\sum y = 14$
$\sum y^2 = 40$
$\sum xy = 882$
$n = 7$
$r = -0.078$

19. continued

H_0: $\rho = 0$

H_1: $\rho \neq 0$

C. V. $= \pm 0.754$ d. f. $= 5$

Decision: do not reject. There is not a significant relationship between age and number of accidents. No regression should be done.

20.

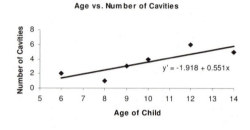

Age vs. Number of Cavities

$y' = -1.918 + 0.551x$

$\sum x = 59$

$\sum x^2 = 621$

$\sum y = 21$

$\sum y^2 = 91$

$\sum xy = 229$

n = 6

r = 0.842

H_0: $\rho = 0$

H_1: $\rho \neq 0$

C. V. $= \pm 0.811$ d. f. $= 4$

Decision: Reject. There is a significant linear relationship between age and number of cavities.

$a = -1.918$ $b = 0.551$

$y' = -1.918 + 0.551x$

When x = 11: $y' = -1.918 + 0.551(11)$

$y' = 4.14$ or 4 cavities

21.

Fat Grams vs. Cholesterol Level

$\sum x = 67.2$

$\sum x^2 = 582.62$

$\sum y = 1740$

$\sum y^2 = 386,636$

$\sum xy = 14847.9$

21. continued

n = 8

r = 0.602

H_0: $\rho = 0$

H_1: $\rho \neq 0$

C. V. $= \pm 0.707$ d. f. $= 6$

Decision: Do not reject. There is no significant linear relationship between fat and cholesterol. No regression should be done.

22.

$S_{est} = \sqrt{\frac{91 - (-1.918)(21) - 0.551(229)}{6-2}}$

$S_{est} = 1.129*$

23.

(For calculation purposes only, since no regression should be done.)

$S_{est} = \sqrt{\frac{386,636 - 110.12(1740) - 12.784(14,847.9)}{8-2}}$

$S_{est} = 29.47*$

24.

$y' = -1.918 + 0.551(7) = 1.939$ or 2

$2 - 2.132(1.129)\sqrt{1 + \frac{1}{6} + \frac{6(11-9.833)^2}{6(621)-59^2}} < y$

$< 2 + 2.132(1.129)\sqrt{1 + \frac{1}{6} + \frac{6(11-9.833)^2}{6(621)-59^2}}$

$2 - 2.132(1.129)(1.095) < y < 2 + 2.132(1.129)(1.095)$

$-0.6 < y < 4.6$ or $0 < y < 5*$

25.

Since no regression should be done, the average of the y' values is used: $\bar{y} = 217.5$

26.

$y' = 98.7 + 3.82(3) + 6.51(1.5) = 119.9*$

27.

$R = \sqrt{\frac{(0.561)^2 + (0.714)^2 - 2(0.561)(0.714)(0.625)}{1 - (0.625)^2}}$

$R = 0.729*$

28.

$R^2_{adj} = 1 - \left[\frac{(1 - 0.774^2)(8-1)}{(8-2-1)} \right]$

$R^2_{adj} = 0.439*$

*These answers may vary due to method of calculation and/or rounding.

Note: Graphs are not to scale and are intended to convey a general idea.

Answers may vary due to rounding, TI-83's, or computer programs.

EXERCISE SET 11-1

1.
The variance test compares a sample variance to a hypothesized population variance, while the goodness of fit test compares a distribution obtained from a sample with a hypothesized distribution.

3.
The expected values are computed based on what the null hypothesis states about the distribution.

5.
H_0: 82% of home-schooled students receive their education entirely at home, 12% attend school up to 9 hours per week, and 6% spend from 9 to 25 hours per week at school.
H_1: The proportions differ from those stated in the null hypothesis. (claim)

C. V. = 5.991   d. f. = 2 $\alpha = 0.05$

O	E
50	0.82(85) = 69.7
25	0.12(85) = 10.2
10	0.06(85) = 5.1

$$\chi^2 = \sum \frac{(O-E)^2}{E} = \frac{(50-69.7)^2}{69.7} + \frac{(25-10.2)^2}{10.2} +$$

$$\frac{(10-5.1)^2}{5.1} = 31.750$$

0   5.991   ↑
                                    31.75

Reject the null hypothesis. There is enough evidence to support the claim that the proportions are different from those stated by the government.

7.
H_0: The distribution of the recorded music sales were as follows: full-length CDs, 77.8%; digital downloads, 12.8%; singles, 3.8%; and other formats, 5.6%.
H_1: The distribution is not the same as that stated in the null hypothesis. (claim)

C. V. = 7.815   d. f. = 3 $\alpha = 0.05$

O	E
48	0.128(200) = 25.6
135	0.778(200) = 155.6
10	0.038(200) = 7.6
7	0.056(200) = 11.2

$$\chi^2 = \sum \frac{(O-E)^2}{E} = \frac{(48-25.6)^2}{25.6} + \frac{(135-155.6)^2}{155.6}$$

$$+ \frac{(10-7.6)^2}{7.6} + \frac{(7-11.2)^2}{11.2} = 24.66$$

0   7.815   ↑
                                        24.66

Reject the null hypothesis. There is enough evidence to support the claim that the distribution is not the same as stated in the null hypothesis.

9.
H_0: The proportions are distributed as follows: safe, 35%; not safe, 52%; no opinion, 13%.
H_1: The distribution is not the same as stated in the null hypothesis. (claim)

C. V. = 9.210   d. f. = 2 $\alpha = 0.01$

O	E
40	0.35(120) = 42
60	0.52(120) = 62.4
20	0.13(120) = 15.6

$$\chi^2 = \frac{(40-42)^2}{42} + \frac{(60-62.4)^2}{62.4} + \frac{(20-15.6)^2}{15.6}$$

$$\chi^2 = 1.4286$$

9. continued

0   ↑   9.210
            1.4286

Do not reject the null hypothesis. There is not enough evidence to support the claim that the proportions are different.

11.
H_0: The distribution of students who use calculators is as follows: never, 28%; sometimes, 51%; and always, 21%.
H_1: The distribution is not the same as stated in the null hypothesis. (claim)

O	E
30	0.28(140) = 39.2
78	0.51(140) = 71.4
32	0.21(140) = 29.4

C. V. = 5.991   d. f. = 2 $\alpha = 0.05$

$\chi^2 = \frac{(30-39.2)^2}{39.2} + \frac{(78-71.4)^2}{71.4} + \frac{(32-29.4)^2}{29.4}$

$\chi^2 = 2.999$

0   ↑   5.991
         2.999

Do not reject the null hypothesis. There is not enough evidence to support the claim that the distribution is different from that stated in the null hypothesis.

13.
H_0: The method of payment for purchases is distributed as follows: 53% cash, 30% checks, 16% credit cards, and 1% no preference. (claim)
H_1: The distribution is not the same as stated in the null hypothesis.

13. continued

O	E
400	0.53(800) = 424
210	0.30(800) = 240
170	0.16(800) = 128
20	0.01(800) = 8

C. V. = 11.345   d. f. = 3 $\alpha = 0.01$

$\chi^2 = \frac{(400-424)^2}{424} + \frac{(210-240)^2}{240} + \frac{(170-128)^2}{128}$

$+ \frac{(20-8)^2}{8} = 36.8897$

0   11.345   ↑
                36.8897

Reject the null hypothesis. There is enough evidence to reject the claim that the distribution is the same as reported in the survey.

15.
H_0: The proportion of Internet users is the same for all groups.
H_1: The proportion of Internet users is not the same for all groups. (claim)

$E = \frac{125}{3} = 41.67$
C. V. = 5.991   d. f. = 2 $\alpha = 0.05$
$\chi^2 = \sum \frac{(O-E)^2}{E} = \frac{(44-41.67)^2}{41.67} + \frac{(41-41.67)^2}{41.67}$

$+ \frac{(40-41.67)^2}{41.67}$

$\chi^2 = 0.208$

0   ↑   5.991
   0.208

Do not reject the null hypothesis. There is not enough evidence to support the claim that the proportions differ.

17.

H_0: The distribution of the ways people pay for their prescriptions is as follows: 60% used personal funds, 25% used insurance, and 15% used Medicare. (claim)

H_1: The distribution is not the same as stated in the null hypothesis.

$\alpha = 0.05$ d. f. $= 2$

P-value > 0.05 (TI: P-value $= 0.7164$)

O	E
32	$0.6(50) = 30$
10	$0.25(50) = 12.5$
8	$0.15(50) = 7.5$

$$\chi^2 = \sum \frac{(O-E)^2}{E} = \frac{(32-30)^2}{30} + \frac{(10-12.5)^2}{12.5} +$$

$$\frac{(8-7.5)^2}{7.5} = 0.667$$

Do not reject the null hypothesis since P-value > 0.05. There is not enough evidence to reject the claim that the distribution is the same as stated in the null hypothesis. An implication of the results is that the majority of people are using their own money to pay for medications. A less expensive medication could help people financially.

19.

Answers will vary.

EXERCISE SET 11-2

1.

The independence test and the goodness of fit test both use the same formula for computing the test-value; however, the independence test uses a contingency table whereas the goodness of fit test does not.

3.

H_0: The variables are independent or not related.

H_1: The variables are dependent or related.

5.

The expected values are computed as (row total · column total) ÷ grand total.

7.

H_0: $p_1 = p_2 = p_3 = \cdots = p_n$

H_1: At least one proportion is different from the others.

9.

H_0: The number of endangered species is independent of the number of threatened species.

H_1: The number of endangered species is dependent upon the number of threatened species. (claim)

C. V. $= 9.488$ d. f. $= 4$ $\alpha = 0.05$

$E = \frac{\text{(row sum)(column sum)}}{\text{grand total}}$

$E_{1,1} = \frac{(247)(81)}{369} = 54.2195$

$E_{1,2} = \frac{(247)(91)}{369} = 60.9133$

$E_{1,3} = \frac{(247)(37)}{369} = 24.7669$

$E_{1,4} = \frac{(247)(23)}{369} = 15.3957$

$E_{1,5} = \frac{(247)(137)}{369} = 91.7046$

$E_{2,1} = \frac{(122)(81)}{369} = 26.7805$

$E_{2,2} = \frac{(122)(91)}{369} = 30.0867$

$E_{2,3} = \frac{(122)(37)}{369} = 12.2331$

$E_{2,4} = \frac{(122)(23)}{369} = 7.6043$

$E_{2,5} = \frac{(122)(137)}{369} = 45.2954$

	Mammal	Bird
Endangered	68(54.2195)	76(60.9133)
Threatened	13(26.7805)	15(30.0867)

	Reptile	Amphibian	Fish
Endangered	14(24.7669)	13(15.3957)	76(91.7046)
Threatened	23(12.2331)	10(7.6043)	61(45.2954)

$$\chi^2 = \frac{(68-54.2195)^2}{54.2195} + \frac{(76-60.9133)^2}{60.9133} + \frac{(13-26.7805)^2}{26.7805}$$

$$+ \frac{(15-30.0867)^2}{30.0867} + \frac{(14-24.7669)^2}{24.7669} + \frac{(13-15.3957)^2}{15.3957}$$

$$+ \frac{(23-12.2331)^2}{12.2331} + \frac{(10-7.6043)^2}{7.6043} + \frac{(76-91.7046)^2}{91.7046}$$

$$+ \frac{(61-45.2954)^2}{45.2954} = 45.3145$$

9. continued

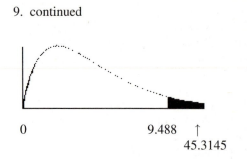

0   9.488   ↑
                   45.3145

Reject the null hypothesis. There is enough evidence to support the claim that there is a relationship between the type of vertebrate and whether it is endangered or threatened. The result is not different for $\alpha = 0.01$.

11.
H_0: The composition of the House of Representatives is independent of the state.
H_1: The composition of the House of Representatives is dependent upon the state. (claim)

C. V. $= 7.815$ d. f. $= 3$ $\alpha = 0.05$
$E = \frac{\text{(row sum)(column sum)}}{\text{grand total}}$

$E_{1,1} = \frac{(203)(320)}{542} = 119.8524$

$E_{1,2} = \frac{(203)(222)}{542} = 83.1476$

$E_{2,1} = \frac{(98)(320)}{542} = 57.8598$

$E_{2,2} = \frac{(98)(222)}{542} = 40.1402$

$E_{3,1} = \frac{(100)(320)}{542} = 59.0406$

$E_{3,2} = \frac{(100)(222)}{542} 40.9594$

$E_{4,1} = \frac{(141)(320)}{542} = 83.2472$

$E_{4,2} = \frac{(141)(222)}{542} = 57.7528$

State	Democrats	Republicans
PA	100(119.8524)	103(83.1476)
OH	39(57.8598)	59(40.1402)
WV	75(59.0406)	25(40.9594)
MD	106(83.2472)	35(57.7528)

$\chi^2 = \sum \frac{(O-E)^2}{E} = \frac{(100-119.8524)^2}{119.8524} + \frac{(103-83.1476)^2}{83.1476}$

$+ \frac{(39-57.8598)^2}{57.8598} + \frac{(59-40.1402)^2}{40.1402} + \frac{(75-59.0406)^2}{59.0406}$

11. continued
$+ \frac{(25-40.9594)^2}{40.9594} + \frac{(106-83.2472)^2}{83.2472} + \frac{(35-57.7528)^2}{57.7528}$

$\chi^2 = 48.7521$

0   7.815   ↑
                      48.7521

Reject the null hypothesis. There is enough evidence to support the claim that the composition of the legislature is dependent upon the state.

13.
H_0: The type of Olympic medal won is independent of the country that won the medal.
H_1: The type of medal won is dependent on the country that won the medal. (claim)

C. V. $= 10.645$ d. f. $= 6$ $\alpha = 0.10$

	Gold	Silver	Bronze
United States	9(13.40)	15(13.72)	13(9.89)
Germany	10(10.86)	13(11.12)	7(8.02)
Canada	14(9.41)	7(9.64)	5(6.95)
Norway	9(8.33)	8(8.53)	6(6.15)

$\chi^2 = \sum \frac{(O-E)^2}{E} = \frac{(9-13.40)^2}{13.40} + \frac{(15-13.72)^2}{13.72}$

$+ \frac{(13-9.89)^2}{9.89} + \frac{(10-10.86)^2}{10.86} + \frac{(13-11.12)^2}{11.12}$

$+ \frac{(7-8.02)^2}{8.02} + \frac{(14-9.41)^2}{9.41} + \frac{(7-9.64)^2}{9.64}$

$+ \frac{(5-6.95)^2}{6.95} + \frac{(9-8.33)^2}{8.33} + \frac{(8-8.53)^2}{8.53}$

$+ \frac{(6-6.15)^2}{6.15}$

$\chi^2 = 6.657$

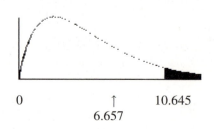

0   ↑   10.645
          6.657

13. continued
Do not reject the null hypothesis. There is not enough evidence to support the claim that the type of medal won is dependent on the country that won the medal.

15.
H_0: The program of study of a student is independent of the type of institution.
H_1: The program of study of a student is dependent upon the type of institution. (claim)

C. V. = 7.815   d. f. = 3 $\alpha = 0.05$

$E_{1,1} = \frac{(88)(302)}{707} = 37.5898$

$E_{1,2} = \frac{(88)(405)}{707} = 50.4102$

$E_{2,1} = \frac{(441)(302)}{707} = 188.3762$

$E_{2,2} = \frac{(441)(405)}{707} = 252.6238$

$E_{3,1} = \frac{(87)(302)}{707} = 37.1627$

$E_{3,2} = \frac{(87)(405)}{707} = 49.8373$

$E_{4,1} = \frac{(91)(302)}{707} = 38.8713$

$E_{4,2} = \frac{(91)(405)}{707} = 52.1287$

	Two-year	Four-year
Agriculture	36(37.5898)	52(50.4102)
Criminal Justice	210(188.3762)	231(252.6238)
Lang/Lit	28(37.1627)	59(49.8373)
Math/Stat	28(38.8713)	63(52.1287)

$\chi^2 = \sum \frac{(O-E)^2}{E} = \frac{(36-37.5898)^2}{37.5898} + \frac{(52-50.4102)^2}{50.4102}$

$+ \frac{(210-188.3762)^2}{188.3762} + \frac{(231-252.6238)^2}{252.6238} + \frac{(28-37.1627)^2}{37.1627}$

$+ \frac{(59-49.8373)^2}{49.8373} + \frac{(28-38.8713)^2}{38.8713} + \frac{(63-52.1287)^2}{52.1287}$

$\chi^2 = 13.702$

0   7.815 ↑
                                              13.702

15. continued
Reject the null hypothesis. There is enough evidence to conclude that the type of program is dependent on the type of institution.

17.
H_0: The type of furniture sold is independent of the store that sold the furniture.
H_1: The type of furniture sold is dependent on the store that sold it. (claim)

C. V. = 9.488   d. f. = 4 $\alpha = 0.05$

	Recliner	Sofa	Loveseat
Store 22A	15(17.31)	12(12.31)	18(15.38)
Store 22B	20(16.15)	10(11.49)	12(14.36)
Store 22C	10(11.54)	10(8.21)	10(10.26)

$\chi^2 = \frac{(15-17.31)^2}{17.31} + \frac{(12-12.31)^2}{12.31} + \frac{(18-15.38)^2}{15.38}$

$+ \frac{(20-16.15)^2}{16.15} + \frac{(10-11.49)^2}{11.49} + \frac{(12-14.36)^2}{14.36}$

$+ \frac{(10-11.54)^2}{11.54} + \frac{(10-8.21)^2}{8.21} + \frac{(10-10.26)^2}{10.26}$

$\chi^2 = 2.864$

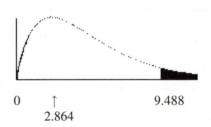

0   ↑   9.488
       2.864

Do not reject the null hypothesis. There is not enough evidence to support the claim that the type of furniture sold is dependent on the store that sold the furniture.

19.
H_0: The choice of exercise equipment is independent of the gender of the individual using it.
H_1: The choice of exercise equipment is dependent upon the gender of the individual using it. (claim)

C. V. = 5.991   d. f. = 2 $\alpha = 0.05$

	Treadmill	Elliptical	Bike
Male	120(105.45)	60(74.30)	75(75.25)
Female	100(114.55)	95(80.70)	82(81.75)

119

19. continued

$$\chi^2 = \sum \frac{(O-E)^2}{E} = \frac{(120-105.45)^2}{105.45} + \frac{(60-74.30)^2}{74.30}$$

$$+ \frac{(75-75.25)^2}{75.25} + \frac{(100-114.55)^2}{114.55} + \frac{(95-80.70)^2}{80.70}$$

$$+ \frac{(82-81.75)^2}{81.75} = 9.139$$

0   5.991   ↑
            9.139

Reject the null hypothesis. There is enough evidence to support the claim that the choice of exercise equipment is dependent upon the gender of the user.

21.
H_0: The type of book purchased by an individual is independent of the gender of the individual. (claim)
H_1: The type of book purchased by an individual is dependent on the gender of the individual.

$\alpha = 0.05$ d. f. = 2

Gender	Mystery	Romance	Self-help
Male	243(214.121)	201(198.260)	191(222.618)
Female	135(163.879)	149(151.740)	202(170.382)

$$\chi^2 = \sum \frac{(O-E)^2}{E} = \frac{(243-214.121)^2}{214.121} + \frac{(201-198.260)^2}{198.260}$$

$$+ \frac{(191-222.618)^2}{222.618} + \frac{(135-163.879)^2}{163.879}$$

$$+ \frac{(149-151.740)^2}{151.740} + \frac{(202-170.382)^2}{170.382} = 19.429$$

P-value < 0.05 (0.00006)
(TI: P-value = 0.00006)
Reject the null hypothesis since P-value < 0.05. There is enough evidence to reject the claim that the type of book purchased is independent of gender.

23.
H_0: $p_1 = p_2 = p_3 = p_4$ (claim)
H_1: At least one proportion is different.

C. V. = 7.815   d. f. = 3

23. continued

$$E(\text{passed}) = \frac{120(167)}{120} = 41.75$$

$$E(\text{failed}) = \frac{120(313)}{120} = 78.25$$

	Southside	West End	East Hills	Jefferson
Passed	49(41.75)	38(41.75)	46(41.75)	34(41.75)
Failed	71(78.25)	82(78.25)	74(78.25)	86(78.25)

$$\chi^2 = \frac{(49-41.75)^2}{41.75} + \frac{(38-41.75)^2}{41.75} + \frac{(46-41.75)^2}{41.75}$$

$$+ \frac{(34-41.75)^2}{41.75} + \frac{(71-78.25)^2}{78.25} + \frac{(82-78.25)^2}{78.25}$$

$$+ \frac{(74-78.25)^2}{78.25} + \frac{(86-78.25)^2}{78.25} = 5.317$$

0   ↑ 7.851
         5.317

Do not reject the null hypothesis. There is not enough evidence to reject the claim that the proportions are equal.

25.
H_0: $p_1 = p_2 = p_3 = p_4$ (claim)
H_1: At least one proportion is different.

C. V. = 7.815   d. f. = 3

	Services	Manufacturing	Government	Other
10 years ago	33(30.6)	13(15)	11(11.4)	3(3)
Now	18(20.4)	12(10)	8(7.6)	2(2)
Total	51	25	19	5

$$\chi^2 = \frac{(33-30.6)^2}{30.6} + \frac{(13-15)^2}{15} + \frac{(11-11.4)^2}{11.4} +$$

$$\frac{(3-3)^2}{3} + \frac{(18-20.4)^2}{20.4} + \frac{(12-10)^2}{10} + \frac{(8-7.6)^2}{7.6}$$

$$+ \frac{(2-2)^2}{2} = 1.172$$

0   ↑   7.815
         1.172

25. continued
Do not reject the null hypothesis. There is not enough evidence to reject the claim that the proportions are the same. Since the survey was done in Pennsylvania, it is doubtful that it can be generalized to the population of the United States.

27.
H_0: $p_1 = p_2 = p_3 = p_4 = p_5$
H_1: At least one proportion is different.

C. V. $= 9.488$ d. f. $= 4$

$E(\text{yes}) = \frac{(104)(75)}{375} = 20.8$

$E(\text{no}) = \frac{(271)(75)}{375} = 54.2$

	18	19	20
Yes	19(20.8)	18(20.8)	23(20.8)
No	56(54.2)	57(54.2)	52(54.2)

	21	22
Yes	31(20.8)	13(20.8)
No	44(54.2)	62(54.2)

$$\chi^2 = \frac{(19-20.8)^2}{20.8} + \frac{(18-20.8)^2}{20.8} + \frac{(23-20.8)^2}{20.8} +$$

$$\frac{(31-20.8)^2}{20.8} + \frac{(13-20.8)^2}{20.8} + \frac{(56-54.2)^2}{54.2} + \frac{(57-54.2)^2}{54.2}$$

$$+ \frac{(52-54.2)^2}{54.2} + \frac{(44-54.2)^2}{54.2} + \frac{(62-54.2)^2}{54.2}$$

$$\chi^2 = 12.028$$

0   9.488 ↑
                           12.028

Reject the null hypothesis. There is enough evidence to conclude that the proportions are different.

29.
H_0: $p_1 = p_2 = p_3 = p_4$ (claim)
H_1: At least one proportion is different.

$\alpha = 0.05$ d. f. $= 3$

$E(\text{on bars}) = \frac{30(62)}{120} = 15.5$

29. continued
$E(\text{not on bars}) = \frac{30(58)}{120} = 14.5$

	N	S	E	W
on	15(15.5)	18(15.5)	13(15.5)	16(15.5)
off	15(14.5)	12(14.5)	17(14.5)	14(14.5)

$$\chi^2 = \frac{(15-15.5)^2}{15.5} + \frac{(18-15.5)^2}{15.5} + \frac{(13-15.5)^2}{15.5} +$$

$$\frac{(16-15.5)^2}{15.5} + \frac{(15-14.5)^2}{14.5} + \frac{(12-14.5)^2}{14.5} +$$

$$\frac{(17-14.5)^2}{14.5} + \frac{(14-14.5)^2}{14.5} = 1.734$$

P-value > 0.10 (0.629)
(TI: P-value $= 0.6291$)

Do not reject the null hypothesis since P-value > 0.05. There is not enough evidence to reject the claim that the proportions are the same.

31.
H_0: $p_1 = p_2 = p_3$ (claim)
H_1: At least one proportion is different.

C. V. $= 4.605$ d. f. $= 2$

$E(\text{list}) = \frac{96(219)}{288} = 73$

$E(\text{no list}) = \frac{96(69)}{288} = 23$

	A	B	C
list	77(73)	74(73)	68(73)
no list	19(23)	22((23)	28(23)

$$\chi^2 = \frac{(77-73)^2}{73} + \frac{(74-73)^2}{73} + \frac{(68-73)^2}{73}$$

$$+ \frac{(19-23)^2}{23} + \frac{(22-23)^2}{23} + \frac{(28-23)^2}{23}$$

$$\chi^2 = 2.401$$

0   ↑   4.605
             2.401

Do not reject the null hypothesis. There is not enough evidence to reject the claim that the proportions are the same.

33.

$$\chi^2 = \frac{(|O-E|-0.5)^2}{E} = \frac{(|12-9.6|-0.5)^2}{9.6}$$

$$+ \frac{(|15-17.4|-0.5)^2}{17.4} + \frac{(|9-11.4|-0.5)^2}{11.4}$$

$$+ \frac{(|23-20.6|-0.5)^2}{20.6}$$

$$= \frac{3.61}{9.6} + \frac{3.61}{17.4} + \frac{3.61}{11.4} + \frac{3.61}{20.6}$$

$$= 0.376 + 0.207 + 0.317 + 0.175 = 1.075$$

REVIEW EXERCISES - CHAPTER 11

1.

H_0: The distribution of trafic fatalities were as follows: used seat belt, 31.58%; did not use seat belt, 59.83%; status unknown, 8.59%.

H_1: The distribution is not as stated in the null hypothesis. (claim)

C. V. = 5.991   d. f. = 2 $\alpha = 0.05$

$$\chi^2 = \sum \frac{(O-E)^2}{E} = \frac{(35-37.896)^2}{37.896} + \frac{(78-71.796)^2}{71.796}$$

$$+ \frac{(7-10.308)^2}{10.308} = 1.819$$

0   ↑   5.991
     1.819

Do not reject the null hypothesis. There is not enough evidence to support the claim that the distribution differs from the one stated in the null hypothesis.

3.

H_0: Opinion is independent of gender.
H_1: Opinion is dependent on gender. (claim)

C. V. = 4.605   d. f. = 2

Gender	Yes	No	Undecided
Men	114(120.968)	30(22.258)	6(6.774)
Women	136(129.032)	16(23.742)	8(7.226)

3. continued

$$\chi^2 = \frac{(114-120.968)^2}{120.968} + \frac{(30-22.258)^2}{22.258} + \frac{(6-6.774)^2}{6.774}$$

$$+ \frac{(136-129.032)^2}{129.032} + \frac{(16-23.742)^2}{23.742} + \frac{(8-7.226)^2}{7.226}$$

$$\chi^2 = 6.166$$

0   4.605   ↑
                              6.166

Reject the null hypothesis. There is enough evidence to support the claim that opinion is dependent on gender.

5.

H_0: The type of investment is independent of the age of the investor.
H_1: The type of investment is dependent upon the age of the investor. (claim)

C. V. = 9.488   d. f. = 4

Age	Large	Small	Inter.
45	20(28.18)	10(15.45)	10(15.45)
65	42(33.82)	24(18.55)	24(18.55)

Age	CD	Bond
45	15(9.55)	45(31.36)
65	6(11.45)	24(37.64)

$$\chi^2 = \frac{(20-28.18)^2}{28.18} + \frac{(10-15.45)^2}{15.45} + \frac{(10-15.45)^2}{15.45}$$

$$+ \frac{(15-9.55)^2}{9.55} + \frac{(45-31.36)^2}{31.36} + \frac{(42-33.82)^2}{33.82} +$$

$$\frac{(24-18.55)^2}{18.55} + \frac{(24-18.55)^2}{18.55} + \frac{(6-11.45)^2}{11.45} +$$

$$\frac{(24-37.64)^2}{37.64} = 27.998 \text{ or } 28.0$$

0   9.488   ↑
                              28.0

5. continued
Reject the null hypothesis. There is enough
evidence to support the claim that the type of
investment is dependent on age.

7.
H_0: $p_1 = p_2 = p_3$ (claim)
H_1: At least one proportion is different.

$\alpha = 0.01$ d. f. = 2

$E(\text{work}) = \frac{80(114)}{240} = 38$

$E(\text{don't work}) = \frac{80(126)}{240} = 42$

	16	17	18
work	45(38)	31(38)	38(38)
don't work	35(42)	49(42)	42(42)

$\chi^2 = \frac{(45-38)^2}{38} + \frac{(31-38)^2}{38} + \frac{(38-38)^2}{38}$

$+ \frac{(35-42)^2}{42} + \frac{(49-42)^2}{42} + \frac{(42-42)^2}{42} = 4.912$

$0.05 < \text{P-value} < 0.10$ (0.086)

Do not reject the null hypothesis since P-
value > 0.01. There is not enough evidence
to reject the claim that the proportions are
the same.

9.
H_0: Health care coverage is independent of
the state of residence of the individual.
H_1: Health care coverage is related to the
state of residence of the individual. (claim)

C. V. = 11.345   d. f. = 3 $\alpha = 0.01$

	With	Without
Arkansas	552(576.639)	123(98.361)
Montana	793(802.169)	146(136.831)
North Dakota	553(524.528)	61(89.472)
Wyoming	447(441.663)	70(75.337)

$\chi^2 = \frac{(552-576.639)^2}{576.639} + \frac{(123-98.361)^2}{98.361}$

$+ \frac{(793-802.169)^2}{802.169} + \frac{(146-136.831)^2}{136.831} + \frac{(553-524.528)^2}{524.528}$

$+ \frac{(61-89.472)^2}{89.472} + \frac{(447-441.663)^2}{441.663} + \frac{(70-75.337)^2}{75.337}$

$\chi^2 = 18.993$

9. continued

0   11.345   ↑
                                18.993

Reject the null hypothesis. There is enough
evidence to support the claim that health
care coverage is related to the state of
residence of the individual.

CHAPTER 11 QUIZ

1. False, it is one-tailed right.
2. True
3. False, there is little agreement between
    observed and expected frequencies.
4. c
5. b
6. d
7. 6
8. Independent
9. Right
10. At least 5

11. H_0: The reasons why people lost their
jobs are equally distributed.   (claim)
H_1: The reasons why people lost their jobs
are not equally distributed.
C. V. = 5.991   d. f. = 2   E = 24
$\chi^2 = \sum \frac{(O-E)^2}{E} = 2.33$
Do not reject the null hypothesis. There is
not enough evidence to reject the claim that
the reasons why people lost their jobs are
equally distributed. The results could have
been different 10 years ago since different
factors of the economy existed then.

12. H_0: Takeout food is consumed
according to the following distribution: at
home - 53%, in the car - 19%, at work -
14%, other - 14%. (claim)
H_1: The distribution is different from that
stated in the null hypothesis.
C. V. = 11.345   d. f. = 3
$\chi^2 = \sum \frac{(O-E)^2}{E} = 5.271$
Do not reject the null hypothesis. There is
not enough evidence to reject the claim that
the distribution is as stated. Fast-food
restaurants may want to make their

12. continued
advertisements appeal to those who like to take their food home to eat.

13. H_0: College students show the same preference for shopping channels as those surveyed.
H_1: College students show a different preference for shopping channels. (claim)
C. V. = 7.815   d. f. = 3
$\chi^2 = 21.789$
Reject the null hypothesis. There is enough evidence to support the claim that college students show a different preference for shopping channels.

14. H_0: The number of commuters is distributed as follows: alone - 75.7%, carpooling - 12.2%, public transportation - 4.7%, walking - 2.9%, other - 1.2%, and working at home - 3.3%.
H_1: The proportions are different from the null hypothesis. (claim)
C. V. = 11.071   d. f. = 5
$\chi^2 = 41.269$
Reject the null hypothesis. There is enough evidence to support the claim that the distribution is different from the one stated in the null hypothesis.

15. H_0: Ice cream flavor is independent of the gender of the purchaser. (claim)
H_1: Ice cream flavor is dependent upon the gender of the purchaser.
C. V. = 7.815   d. f. = 3
$\chi^2 = 7.198$
Do not reject the null hypothesis. There is not enough evidence to reject the claim that ice cream flavor is independent of the gender of the purchaser.

16. H_0: The type of pizza ordered is independent of the age of the purchaser.
H_1: The type of pizza ordered is dependent on the age of the purchaser. (claim)
$\alpha = 0.10$ d. f. = 9
$\chi^2 = 107.3$
P-value < 0.005
Reject the null hypothesis since P-value < 0.10. There is enough evidence to support the claim that the type of pizza is related to the age of the purchaser.

17. H_0: The color of the pennant purchased is independent of the gender of the purchaser. (claim)
H_1: The color of the pennant purchased is dependent on the gender of the purchaser.
C. V. = 4.605   d. f. = 2
$\chi^2 = 5.632$
Reject the null hypothesis. There is enough evidence to reject the claim that the color of the pennant purchased is independent of the gender of the purchaser.

18. H_0: The opinion of the children on the use of the tax credit is independent of the gender of the children.
H_1: The opinion of the children on the use of the tax credit is dependent upon the gender of the children. (claim)
C. V. = 4.605   d. f. = 2
$\chi^2 = 1.534$
Do not reject the null hypothesis. There is not enough evidence to support the claim that the opinion of the children is dependent upon their gender.

19. H_0: $p_1 = p_2 = p_3$ (claim)
H_1: At least one proportion is different from the others.
C. V. = 4.605   d. f. = 2
$\chi^2 = 6.711$
Reject the null hypothesis. There is enough evidence to reject the claim that the proportions are equal. It seems that more women are undecided about their jobs. Perhaps they want better income or greater chances of advancement.

Note: Graphs are not to scale and are intended to convey a general idea. Answers may vary due to rounding.

EXERCISE SET 12-1

1.

The analysis of variance using the F-test can be used to compare 3 or more means.

3.

The populations from which the samples were obtained must be normally distributed. The samples must be independent of each other. The variances of the populations must be equal.

5.

$$F = \frac{s_B^2}{s_W^2}$$

7.

Scheffe´ Test and the Tukey Test

9.

H_0: $\mu_1 = \mu_2 = \mu_3$
H_1: At least one mean is different from the others. (claim)

C. V. = 4.26 $\alpha = 0.05$
d. f. N. = 2   d. f. D. = 9
$\overline{X}_1 = 1.888$ $s_1 = 0.535$
$\overline{X}_2 = 2.224$ $s_2 = 0.1328$
$\overline{X}_3 = 3.525$ $s_3 = 0.1344$
$\overline{X}_{GM} = \frac{27.61}{12} = 2.301$

$$s_B^2 = \frac{5(1.888-2.301)^2+5(2.224-2.301)^2+2(3.525-2.301)^2}{3-1}$$

$$s_B^2 = 1.9394$$

$$s_W^2 = \frac{4(0.535)^2+4(0.1328)^2+1(0.1344)^2}{4+4+1}$$

$$s_W^2 = 0.1371$$

$$F = \frac{1.9394}{0.1371} = 14.146 \quad (TI: \ F = 14.1489)$$

9. continued
Reject the null hypothesis. There is enough evidence to conclude that at least one mean is different from the others.

11.
H_0: $\mu_1 = \mu_2 = \mu_3$
H_1: At least one mean is different from the others. (claim)

C. V. = 3.98 $\alpha = 0.05$
d. f. N = 2   d. f. D = 11
$\overline{X}_{GM} = \frac{52414}{14} = 3743.857$

$$s_B^2 = 3{,}633{,}540.88$$

$$s_W^2 = 1{,}330{,}350$$

$$F = \frac{3633540.88}{1330350} = 2.7313$$

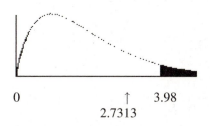

Do not reject the null hypothesis. There is not enough evidence to support the claim that at least one mean is different from the others.

13.
H_0: $\mu_1 = \mu_2 = \mu_3$
H_1: At least one mean is different. (claim)

k = 3   N = 18   d.f.N. = 2   d.f.D. = 15
CV = 3.68
$\overline{X}_1 = 7$ $s_1^2 = 1.37$
$\overline{X}_2 = 8.12$ $s_2^2 = 0.64$
$\overline{X}_3 = 5.23$ $s_3^2 = 2.66$
$\overline{X}_{GM} = 6.7833$

$$s_B^2 = \frac{6(7-6.78)^2}{2} + \frac{6(8.12-6.78)^2}{2}$$

$$+ \frac{6(5.23-6.78)^2}{2} = 12.7$$

$$s_W^2 = \frac{5(1.37)+5(0.64)+5(2.66)}{5+5+5}$$

$$s_W^2 = \frac{23.35}{15} = 1.56$$

$$F = \frac{12.7}{1.56} = 8.14$$

13. continued

0   3.68   ↑
                    8.14

Reject the null hypothesis. There is enough evidence to support the claim that at least one mean is different.

15.
H_0: $\mu_1 = \mu_2 = \mu_3$ (claim)
H_1: At least one mean is different from the others.

C. V. = 4.10 $\alpha = 0.10$
d. f. N = 2   d. f. D = 10
$\overline{X}_1 = 35.4$ $s_1^2 = 351.8$
$\overline{X}_2 = 68.75$ $s_2^2 = 338.25$
$\overline{X}_3 = 44.25$ $s_3^2 = 277.583$
$\overline{X}_{GM} = \frac{629}{13} = 48.385$

$$s_B^2 = \frac{\sum n_i (\overline{X}_i - \overline{X}_{GM})^2}{k-1}$$

$$s_B^2 = \frac{5(35.4 - 48.385)^2}{2} + \frac{4(68.75 - 48.385)^2}{2}$$

$$+ \frac{4(44.25 - 48.385)^2}{2} = 1285.188$$

$$s_W^2 = \frac{\sum (n_i - 1) s_i^2}{\sum (n_i - 1)}$$

$$s_W^2 = \frac{4(351.8) + 3(338.25) + 3(277.583)}{4 + 3 + 3}$$

$s_W^2 = 325.47$
$F = \frac{s_B^2}{s_W^2} = \frac{1285.188}{325.47} = 3.9487$

Do not reject the null hypothesis. There is enough evidence to reject the claim that the means are the same.

17.
H_0: $\mu_1 = \mu_2 = \mu_3$
H_1: At least one mean is different from the others. (claim)

C. V. = 2.61 $\alpha = 0.10$
d. f. N = 2   d. f. D = 19
$\overline{X}_1 = 233.33$ $s_1 = 28.225$
$\overline{X}_2 = 203.125$ $s_2 = 39.364$

17. continued
$\overline{X}_3 = 155.625$ $s_3 = 28.213$
$\overline{X}_{GM} = 194.091$

$$s_B^2 = \frac{21,729.735}{2} = 10,864.8675$$

$$s_W^2 = \frac{20,402.083}{19} = 1073.794$$

$$F = \frac{s_B^2}{s_W^2} = \frac{10,864.8675}{1073.794} = 10.12$$

P-value = 0.00102
Reject since P-value < 0.10.

19.
H_0: $\mu_1 = \mu_2 = \mu_3$
H_1: At least one mean differs from the others. (claim)

C. V. = 2.57 $\alpha = 0.05$
d. f. N = 2   d. f. D = 21
$\overline{X}_{GM} = \frac{1748}{24} = 72.83$
$\overline{X}_1 = 78.625$ $s_1 = 8.568$
$\overline{X}_2 = 69.875$ $s_2 = 8.823$
$\overline{X}_3 = 70$ $s_3 = 4.629$

$$s_B^2 = \frac{402.583}{2} = 201.292$$

$$s_W^2 = \frac{1208.75}{21} = 57.560$$

$$F = \frac{201.292}{57.560} = 3.497$$

0   2.52   ↑
                    3.497

Reject the null hypothesis. There is enough evidence to support the claim that at least one mean is different.

EXERCISE SET 12-2

1.
The Scheffé and Tukey tests are used.

3.
Scheffé Test
C. V. = 8.52

3. continued

$$F_s = \frac{(\overline{X}_i - \overline{X}_j)^2}{s_W^2\left(\frac{1}{n_i} + \frac{1}{n_j}\right)}$$

For $\overline{X}_1$ vs $\overline{X}_2$

$$F_S = \frac{(1.888 - 2.224)^2}{0.13707\left(\frac{1}{5} + \frac{1}{5}\right)} = \frac{0.112896}{0.054828} = 2.059$$

For $\overline{X}_1$ vs $\overline{X}_3$

$$F_S = \frac{(1.888 - 3.525)^2}{0.13707\left(\frac{1}{5} + \frac{1}{2}\right)} = \frac{2.679769}{0.095949} = 27.929$$

For $\overline{X}_2$ vs $\overline{X}_3$

$$F_S = \frac{(2.224 - 3.525)^2}{0.13707\left(\frac{1}{5} + \frac{1}{2}\right)} = \frac{1.692601}{0.0959504} = 17.640$$

There is a significant difference between $\overline{X}_1$ and $\overline{X}_3$ and between $\overline{X}_2$ and $\overline{X}_3$.

5.
Tukey Test:
C. V. = 3.29

$\overline{X}_1$ vs $\overline{X}_2$:
$$q = \frac{7 - 8.12}{\sqrt{\frac{1.56}{6}}} = -2.196$$

$\overline{X}_1$ vs $\overline{X}_3$:
$$q = \frac{7 - 5.23}{\sqrt{\frac{1.56}{6}}} = 3.47$$

$\overline{X}_2$ vs $\overline{X}_3$:
$$q = \frac{8.12 - 5.23}{\sqrt{\frac{1.56}{6}}} = 5.667$$

There is a significant difference between $\overline{X}_1$ and $\overline{X}_3$ and between $\overline{X}_2$ and $\overline{X}_3$.

7.
Scheffé Test
C. V. = 5.22

$\overline{X}_1$ vs $\overline{X}_2$:
$$F_s = \frac{(\overline{X}_i - \overline{X}_j)^2}{s_W^2\left(\frac{1}{n_i} + \frac{1}{n_j}\right)} = \frac{(233.33 - 203.125)^2}{1073.776\left(\frac{1}{6} + \frac{1}{8}\right)}$$
$$F_s = 2.91$$

$\overline{X}_1$ vs $\overline{X}_3$:
$$F_s = \frac{(233.33 - 155.625)^2}{1073.776\left(\frac{1}{6} + \frac{1}{8}\right)} = 19.28$$

$\overline{X}_2$ vs $\overline{X}_3$:
$$F_s = \frac{(203.125 - 155.625)^2}{1073.776\left(\frac{1}{8} + \frac{1}{8}\right)} = 8.40$$

There is a significant difference between $\overline{X}_1$ and $\overline{X}_3$ and between $\overline{X}_2$ and $\overline{X}_3$.

9.
Tukey Test:
C. V. = 3.08

$$q = \frac{\overline{X}_i - \overline{X}_j}{\sqrt{\frac{s_W^2}{n}}}$$

$\overline{X}_1$ vs $\overline{X}_2$:
$$q = \frac{(78.625 - 69.875)}{\sqrt{\frac{57.5595}{8}}} = 3.262$$

$\overline{X}_1$ vs $\overline{X}_3$:
$$q = \frac{(78.625 - 70)}{\sqrt{\frac{57.5595}{8}}} = 3.215$$

$\overline{X}_2$ vs $\overline{X}_3$:
$$q = \frac{(69.875 - 70)}{\sqrt{\frac{57.5595}{8}}} = -0.047$$

There is a significant difference between $\overline{X}_1$ and $\overline{X}_2$ and $\overline{X}_1$ and $\overline{X}_3$.

11.
H_0: $\mu_1 = \mu_2 = \mu_3$
H_1: At least one mean is different from the others. (claim)
C. V. = 3.47 $\alpha = 0.05$
d. f. N. = 2   d. f. D. = 21
$\overline{X}_{GM} = 4.554$ $s_B^2 = 9.82113$ $s_W^2 = 4.93225$

$$F = \frac{9.82113}{4.93225} = 1.9912$$

0   ↑   3.47
                  1.9912

Do not reject the null hypothesis. There is not enough evidence to support the claim that at least one mean is different from the others.

13.
H_0: $\mu_1 = \mu_2 = \mu_3$
H_1: At least one mean is different. (claim)

C. V. = 3.68 $\alpha = 0.05$
d. f. N = 2   d. f. D = 15
$\overline{X}_{GM} = \frac{4566}{18} = 253.67$

$$s_B^2 = \frac{39,374.11}{2} = 19,687.05555$$

13. continued
$$s_w^2 = \frac{17{,}197}{15} = 1146.46667$$

$$F = \frac{19{,}687.05555}{1146.46667} = 17.172$$

| 0 | 3.68 | ↑ |
| | | 17.172 |

Reject the null hypothesis. There is enough evidence to support the claim that at least one mean is different.

Tukey Test:
C. V. = 3.67

$\overline{X}_1$ vs $\overline{X}_2$:
$$q = \frac{(208.17 - 321.17)}{\sqrt{\frac{1146.467}{6}}} = -8.17$$

$\overline{X}_1$ vs $\overline{X}_3$:
$$q = \frac{(208.17 - 248.33)}{\sqrt{\frac{1146.467}{6}}} = -2.91$$

$\overline{X}_2$ vs $\overline{X}_3$:
$$q = \frac{(321.17 - 248.33)}{\sqrt{\frac{1146.467}{6}}} = 5.27$$

There is a significant difference between $\overline{X}_1$ and $\overline{X}_2$ and between $\overline{X}_2$ and $\overline{X}_3$.

EXERCISE SET 12-3

1.
The two-way ANOVA allows the researcher to test the effects of two independent variables and a possible interaction effect. The one-way ANOVA can test the effects of one independent variable only.

3.
The mean square values are computed by dividing the sum of squares by the corresponding degrees of freedom.

5.
a. d. f.$_A$ = (3 − 1) = 2 for factor A
b. d. f.$_B$ = (2 − 1) = 1 for factor B
c. d. f.$_{AxB}$ = (3 − 1)(2 − 1) = 2
d. d. f.$_{within}$ = 3 · 2(5 − 1) = 24

7.
The two types of interactions that can occur are ordinal and disordinal.

9.
a. The lines will be parallel or approximately parallel. They could also coincide.
b. The lines will not intersect and they will not be parallel.
c. The lines will intersect.

11.
H_0: There is no interaction effect between temperature and level of humidity.
H_1: There is an interactive effect between temperature and level of humidity.

H_0: There is no difference in mean length of effectiveness with respect to humidity.
H_1: There is a difference in mean length of effectiveness with respect to humidity.

H_0: There is no difference in mean length of effectiveness based on temperature.
H_1: There is a difference in mean length of effectiveness based on temperature.

ANOVA SUMMARY TABLE

Source	SS	d. f.	MS	F
Humidity	280.3333	1	280.3333	18.3825
Temperature	3	1	3	0.1967
Interaction	65.3333	1	65.3333	4.2842
Within	122	8	15.25	
Total	470.6667	11		

The critical value at $\alpha = 0.05$ with d. f. N = 1 and d. f. D = 8 is 5.318 for F_A, F_B, and F_{AxB}.

Since the only F test value that exceeds the critical value is the one for humidity, there is sufficient evidence to conclude that there is a difference in mean length of effectiveness based on the humidity level. The temperature and interaction effects are not significant.

| 0 | ↑ | ↑ | 5.318 | ↑ |
| | 0.1967 | 4.2842 | | 18.3825 |

13.
Interaction:
H_0: There is no interaction effect on the durability rating between the dry additives and the solution-based additives.
H_1: There is an interaction effect on the durability rating between the dry additives and the solution-based additives.

Solution-based Additive:
H_0: There is no difference in the mean durability rating with respect to the solution-based additives.
H_1: There is a difference in the mean durability rating with respect to the solution-based additives.

Dry Additives:
H_0: There is no difference in the mean durability rating with respect to the dry additive.
H_1: There is a difference in the mean durability rating with respect to the dry additive.

ANOVA SUMMARY TABLE

Source	SS	d. f.	MS	F
Solution	1.563	1	1.563	0.497
Dry	0.063	1	0.063	0.020
Interaction	1.563	1	1.563	0.497
Within	37.750	12	3.146	
Total	40.939	15		

The critical value at $\alpha = 0.05$ with d. f. N = 1 and d. f. D = 12 is 4.75. F = 0.497 for the solution-based additive and F = 0.020 for the dry additive. There is not enough evidence to conclude an effect on the mean durability based on the additives. For interaction, there is also not a significant interaction effect.

15.
H_0: There is no interaction effect between the ages of the salespersons and the products they sell on the monthly sales.
H_1: There is an interaction effect between the ages of the salespersons and the products they sell on the monthly sales.

H_0: There is no difference in the means of the monthly sales of the two age groups.
H_1: There is a difference in the means of the monthly sales of the two age groups.

15. continued
H_0: There is no difference among the means of the sales for the different products.
H_1: There is a difference among the means of the sales for the different products.

ANOVA SUMMARY TABLE

Source	SS	d. f.	MS	F
Age	168.033	1	168.033	1.567
Product	1762.067	2	881.034	8.215
Interaction	7955.267	2	3977.634	37.087
Error	2574.000	24	107.250	
Total	12459.367	29		

At $\alpha = 0.05$, the critical values are:

For age, d. f. N = 1, d. f. D = 24, C. V. = 4.26

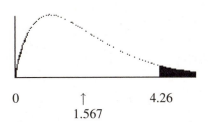

0   ↑   4.26
        1.567

For product and interaction, d. f. N = 2, d. f. D = 24, and C. V. = 3.40

0   3.40 ↑   ↑
                        8.215 37.087

The null hypotheses for the interaction effect and for the type of product sold are rejected since the F test values exceed the critical value, 3.40. The cell means are:

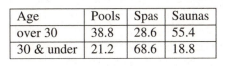

Age	Pools	Spas	Saunas
over 30	38.8	28.6	55.4
30 & under	21.2	68.6	18.8

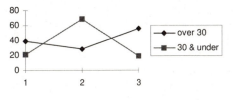

15. continued

Since the lines cross, there is a disordinal interaction hence there is an interaction effect between the age of the sales person and the type of products sold.

REVIEW EXERCISES - CHAPTER 12

1.

H_0: $\mu_1 = \mu_2 = \mu_3$ (claim)

H_1: At least one mean is different from the others.

C. V. = 5.39 $\alpha = 0.01$

d. f. N = 2   d. f. D = 33

$\overline{X}_1 = 620.5$ $s_1^2 = 5445.91$

$\overline{X}_2 = 610.17$ $s_2^2 = 22{,}108.7$

$\overline{X}_3 = 477.83$ $s_3^2 = 5280.33$

$\overline{X}_{GM} = \frac{20{,}502}{36} = 569.5$

$s_B^2 = \frac{151{,}890.667}{2} = 75{,}945.333$

$s_W^2 = \frac{361{,}184.333}{33} = 10{,}944.9798$

$F = \frac{s_B^2}{s_W^2} = \frac{75{,}945.333}{10{,}944.9798} = 6.94$

0   5.39   ↑ 6.94

Reject. At least one mean is different.

Tukey Test   C. V. = 4.45 using (3, 33)

$\overline{X}_1$ vs $\overline{X}_2$:

$q = \frac{\overline{X}_1 - \overline{X}_2}{\sqrt{\frac{s_W^2}{n}}} = \frac{620.5 - 610.17}{\sqrt{\frac{10{,}944.98}{12}}} = 0.342$

$\overline{X}_1$ vs $\overline{X}_3$:

$q = \frac{620.5 - 477.83}{\sqrt{\frac{10{,}944.98}{12}}} = 4.72$

$\overline{X}_2$ vs $\overline{X}_3$:

$q = \frac{610.17 - 477.83}{\sqrt{\frac{10{,}944.98}{12}}} = 4.38$

There is a significant difference between $\overline{X}_1$ and $\overline{X}_3$.

3.

H_0: $\mu_1 = \mu_2 = \mu_3$

H_1: At least one mean is different. (claim)

3. continued

C. V. = 3.55 $\alpha = 0.05$

d. f. N = 2   d. f. D = 18

$\overline{X}_1 = 29.625$ $s_1^2 = 59.125$

$\overline{X}_2 = 29$ $s_2^2 = 63.333$

$\overline{X}_3 = 28.5$ $s_3^2 = 37.1$

$\overline{X}_{GM} = 29.095$

$s_B^2 = \frac{\sum n_i (\overline{X}_i - \overline{X}_{GM})^2}{k - 1}$

$s_B^2 = \frac{8(29.625 - 29.095)^2}{2} + \frac{7(29 - 29.095)^2}{2}$

$+ \frac{6(28.5 - 29.095)^2}{2} = 2.21726$

$s_W^2 = \frac{\sum (n_i - 1) s_i^2}{\sum (n_i - 1)}$

$s_W^2 = \frac{7(59.125) + 6(63.333) + 5(37.1)}{7 + 6 + 5}$

$s_W^2 = 54.509611$

$F = \frac{s_B^2}{s_W^2} = \frac{2.21726}{54.509611} = 0.04075$

Do not reject the null hypothesis. There is not enough evidence to support the claim that at least one mean is different from the others.

5.

H_0: $\mu_1 = \mu_2 = \mu_3$

H_1: At least one mean is different. (claim)

C. V. = 2.61 $\alpha = 0.10$

d. f. N = 2   d. f. D = 19

$\overline{X}_{GM} = 3.8591$

$s_B^2 = 1.65936$

$s_W^2 = 3.40287$

$F = \frac{1.65936}{3.40287} = 0.4876$

0   ↑   2.61
        0.4876

Do not reject. There is not enough evidence to support the claim that at least one mean is different from the others.

7.

H_0: $\mu_1 = \mu_2 = \mu_3 = \mu_4$

H_1: At least one mean is different from the others. (claim)

C. V. = 3.59 $\alpha = 0.05$

d. f. N = 3   d. f. D = 11

$\overline{X}_{GM} = 12.267$

$s_B^2 = 21.422$

$s_W^2 = 117.697$

$F = \frac{21.422}{117.697} = 0.182$

0   ↑   3.59
     0.182

Do not reject the null hypothesis. There is not enough evidence to support the claim that at least one mean is different from the others.

9.

H_0: There is no interaction effect between the type of exercise program and the type of diet on a person's glucose level.

H_1: There is an interaction effect between the type of exercise program and the type of diet on a person's glucose level.

H_0: There is no difference in the means for the glucose levels of the persons in the two exercise programs.

H_1: There is a difference in the means for the glucose levels of the persons in the two exercise programs.

H_0: There is no difference in the means for the glucose levels of the persons in the two diet programs.

H_1: There is a difference in the means for the glucose levels of the persons in the two diet programs.

9. continued

ANOVA SUMMARY TABLE

Source	SS	d. f.	MS	F
Exercise	816.750	1	816.750	60.50
Diet	102.083	1	102.083	7.56
Interaction	444.083	1	444.083	32.90
Within	108.000	8	13.500	
Total	1470.916	11		

At $\alpha = 0.05$ and d. f. N = 1 and d. f. D = 8 the critical value is 5.32 for each F_A, F_B, and $F_{A \times B}$.

Hence all three null hypotheses are rejected.

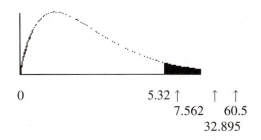

0   5.32 ↑   ↑   ↑
                        7.562   60.5
                                32.895

The cell means should be calculated.

Exercise	Diet A	Diet B
I	64.000	57.667
II	68.333	86.333

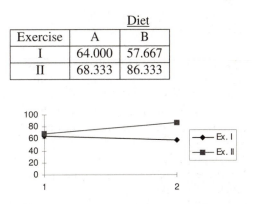

Since the means for the Exercise Program I are both smaller than those for Exercise Program II and the vertical differences are not the same, the interaction is ordinal. Hence one can say that there is a difference for exercise, diet; and that an interaction effect is present.

CHAPTER 12 QUIZ

1. False, there could be a significant difference between only some of the means.
2. False, degrees of freedom are used to find the critical value.
3. False, the null hypothesis should not be rejected.

4. True
5. d
6. a
7. a
8. c
9. ANOVA
10. Tukey
11. Two

12. H_0: $\mu_1 = \mu_2 = \mu_3 = \mu_4$
H_1: At least one mean is different from the others. (claim)
C. V. = 3.49

$s_B^2 = 116.646$ $s_W^2 = 36.132$
$F = \frac{116.646}{36.132} = 3.23$

Do not reject the null hypothesis. There is not enough evidence to support the claim that the means are different.

13. H_0: $\mu_1 = \mu_2 = \mu_3$
H_1: At least one mean is different from the others. (claim)
C. V. = 6.93 $\alpha = 0.01$
$s_B^2 = 119.467$ $s_W^2 = 34.167$
$F = \frac{119.467}{34.167} = 3.497$

Do not reject the null hypothesis. There is not enough evidence to support the claim that at least one mean is different from the others. Writers would want to target their material to the age group of the viewers.

14. H_0: $\mu_1 = \mu_2 = \mu_3$
H_1: At least one mean is different from the others. (claim)
C. V. = 4.26 $\alpha = 0.05$
$s_B^2 = 9.6658$ $s_W^2 = 0.9642$
$F = \frac{9.6658}{0.9642} = 10.025$
Reject the null hypothesis. There is enough evidence to support the claim that at least one mean is different from the others.

Tukey Test:
C. V. = 3.95
$\overline{X}_1 = 7.015$
$\overline{X}_2 = 7.64$
$\overline{X}_3 = 4.69$
$\overline{X}_1$ vs $\overline{X}_2$: q = -1.28
$\overline{X}_1$ vs $\overline{X}_3$: q = 4.74
$\overline{X}_2$ vs $\overline{X}_3$: q = 6.02
There is a significant difference between $\overline{X}_1$ and $\overline{X}_3$ and between $\overline{X}_2$ and $\overline{X}_3$.

15. H_0: $\mu_1 = \mu_2 = \mu_3$ (claim)
H_1: At least one mean is different from the others.
C. V. = 2.92 $\alpha = 0.05$
$s_B^2 = 2114.985$ $s_W^2 = 317.958$
$F = \frac{2114.985}{317.958} = 6.652$
Reject the null hypothesis. There is enough evidence to support the claim that the means are not the same.

Scheffe´ Test:
C. V. = 8.918
For $\overline{X}_1$ vs $\overline{X}_2$, $F_S = 9.32$
For $\overline{X}_1$ vs $\overline{X}_3$, $F_S = 10.124$
For $\overline{X}_2$ vs $\overline{X}_3$, $F_S = 0.126$
There is a significant difference between $\overline{X}_1$ and $\overline{X}_2$ and between $\overline{X}_1$ and $\overline{X}_3$.

16. H_0: $\mu_1 = \mu_2 = \mu_3 = \mu_4$
H_1: At least one mean is different from the others. (claim)
C. V. = 3.07 $\alpha = 0.05$
$s_B^2 = 15.3016$ $s_W^2 = 33.5283$
$F = \frac{15.3016}{33.5283} = 0.4564$
Do not reject. There is not enough evidence to support the claim that at least one mean is different.

17.
a. two-way ANOVA
b. diet and exercise program
c. 2

d. H_0: There is no interaction effect between the type of exercise program and the type of diet on a person's weight loss.
H_1: There is an interaction effect between the type of exercise program and the type of diet on a person's weight loss.

H_0: There is no difference in the means of the weight losses for those in the exercise programs.
H_1: There is a difference in the means of the weight losses for those in the exercise programs.

H_0: There is no difference in the means of the weight losses for those in the diet programs.
H_1: There is a difference in the means of the weight losses for those in the diet programs.

17. continued
e. Diet: F = 21.0, significant
Exercise Program: F = 0.429, not significant
Interaction: F = 0.429, not significant

f. Reject the null hypothesis for the diets.

Note: Graphs are not to scale and are intended to convey a general idea. Answers may vary due to rounding.

EXERCISE SET 13-1

1.
Non-parametric means hypotheses other than those using population parameters can be tested, whereas distribution free means no assumptions about the population distributions have to be satisfied.

3.
The advantages of non-parametric methods are:
1.   They can be used to test population parameters when the variable is not normally distributed.
2.   They can be used when data is nominal or ordinal in nature.
3.   They can be used to test hypotheses other than those involving population parameters.
4.   The computations are easier in some cases than the computations of the parametric counterparts.
5.   They are easier to understand.
The disadvantages are:
1.   They are less sensitive than their parametric counterparts.
2.   They tend to use less information than their parametric counterparts.
3.   They are less efficient than their para-metric counterparts.

5.
DATA	22	32	34	43	43	65	66	71
RANK	1	2	3	4.5	4.5	6	7	8

7.
DATA	3.2	5.9	10.3	11.1	19.4	21.8	23.1
RANK	1	2	3	4	5	6	7

9.
DATA	11	28	36	41	47	50	50	50	52	71	71	88
RANK	1	2	3	4	5	7	7	7	9	10.5	10.5	12

EXERCISE SET 13-2

1.
The sign test uses only + or − signs.

3.
The smaller number of + or − signs

5.

+	+	+	−
+	+	+	+
+	+	−	+
−	−	+	−

H_0: Median = 27.6
H_1: Median $\neq$ 27.6 (claim)

$\alpha = 0.05$ n = 16
C. V. = 3
Test value = 5

Since $5 > 3$, do not reject the null hypothesis. There is not enough evidence to support the claim that the median is not 27.6.

7.

−	+	+	−	−
+	+	+	−	−
−	−	+	−	−
−	+	−	0	−

H_0: Median = 25   (claim)
H_1: Median $\neq$ 25

$\alpha = 0.05$ n = 19
C. V. = 4   Test value = 7

Since $7 > 4$, do not reject the null hypothesis. There is not enough evidence to reject the claim that the median is 25. School boards could use the median to plan for the costs of cyber school enrollments.

9.
H_0: Median = \$10.86   (claim)
H_1: Median $\neq$ \$10.86

C. V. = ± 1.96

$$z = \frac{(x + 0.5) - \left(\frac{n}{2}\right)}{\frac{\sqrt{n}}{2}} = \frac{(18 + 0.5) - \frac{42}{2}}{\frac{\sqrt{42}}{2}}$$

$$z = \frac{-2.5}{3.24} = -0.77$$

−1.96   ↑   0   1.96
     −0.77

9. continued
Do not reject the null hypothesis. There is not enough evidence to reject the claim that the median is $10.86. Home buyers could estimate the yearly cost of their gas bills.

11.
H_0: Median $= 150$ (claim)
H_1: Median $\neq 150$

$$z = \frac{(x + 0.5) - \left(\frac{n}{2}\right)}{\frac{\sqrt{n}}{2}} = \frac{(9 + 0.5) - \frac{35}{2}}{\frac{\sqrt{35}}{2}}$$

$z = -2.70$

C. V. $= \pm 1.96$

$\uparrow \; -1.96 \qquad 0 \qquad 1.96$
-2.70

Reject. There is not enough evidence to support the claim that the median number of faculty members is 150.

13.
H_0: Median $= 50$ (claim)
H_1: Median $\neq 50$

$$z = \frac{(x + 0.5) - \left(\frac{n}{2}\right)}{\frac{\sqrt{n}}{2}} = \frac{(38 + 0.5) - \frac{100}{2}}{\frac{\sqrt{100}}{2}}$$

$z = -2.3$

P-value $= 0.0214$

Reject. There is enough evidence to reject the claim that 50% of the students are against extending the school year.

15.

A	B	C	D	E	F	G	H
+	+	+	+	+	−	+	+

H_0: The medication has no effect on weight loss.
H_1: The medication affects weight loss. (claim)
$\alpha = 0.05$ n = 8
C. V. $= 0$ Test value $= 1$

15. continued
Since $1 > 0$, do not reject the null hypothesis. There is not enough evidence to support the claim that the medication affects weight loss.

17.

1	2	3	4	5	6	7	8
+	+	+	−	+	+	+	+

H_0: There is no difference in test scores.
H_1: There is a difference in test scores.

$\alpha = 0.05$ n = 8
C. V. $= 0$ Test value $= 2$

Since $2 > 0$, do not reject the null hypothesis. There is not enough evidence to support the claim that there is a difference in test scores.

19.

1	2	3	4	5	6	7	8	9	10
+	+	+	+	+	+	−	+	+	−

H_0: The number of viewers is the same as last year. (claim)
H_1: The number of viewers is not the same as last year.

$\alpha = 0.01$ n = 10
C. V. $= 0$ Test value $= 2$

Since $2 > 0$, do not reject the null hypothesis. There is not enough evidence to reject the claim that the number of viewers is the same as last year.

21.
3, 4, 6, 9, 12, 15, 15, 16, 18, 22, 25, 30
At $\alpha = 0.05$, the value from Table J with n = 12 is 2; hence, count in 3 numbers from each end to get $6 \leq MD \leq 22$.

23.
4.2, 4.5, 4.7, 4.8, 5.1, 5.2, 5.6, 6.3, 7.1, 7.2, 7.8, 8.2, 9.3, 9.3, 9.5, 9.6

At $\alpha = 0.02$, the value from Table J with n = 16 is 2; hence, count 3 numbers from each end to get $4.7 \leq MD \leq 9.3$.

25.
12, 14, 14, 15, 16, 17, 18, 19, 19, 21, 23, 25, 27, 32, 33, 35, 39, 41, 42, 47

25. continued

At $\alpha = 0.05$, the value from Table J with n = 20 is 5; hence, count in 6 numbers from each end to get $17 \leq MD \leq 33$.

EXERCISE SET 13-3

1.

The sample sizes must be greater than or equal to 10.

3.

The standard normal distribution

5.

H_0: There is no difference in scores.
H_1: There is a difference in scores between the two groups. (claim)

C. V. = ± 1.96 $\alpha = 0.05$

29	30	32	32	33	35	35	36	36
1	2	3.5	3.5	5	6.5	6.5	8.5	8.5
F	F	F	F	T	F	T	T	T

37	38	38	39	39	40	40	41	43
10	11.5	11.5	13.5	13.5	15.5	15.5	17	18
T	F	T	F	F	F	F	T	T

44	45	46	47
19	20	21	22
T	T	T	T

R = 108 (for freshmen)

$\mu_R = \frac{n_1(n_1 + n_2 + 1)}{2}$

$\mu_R = \frac{11(11 + 11 + 1)}{2} = \frac{11(23)}{2} = \frac{253}{2} = 126.5$

$\sigma_R = \sqrt{\frac{n_1 \cdot n_2(n_1 + n_2 + 1)}{12}}$

$\sigma_R = \sqrt{\frac{11 \cdot 11(11 + 11 + 1)}{12}} = \sqrt{\frac{(11)(11)(23)}{12}}$

$\sigma_R = \sqrt{231.917} = 15.23$

$Z = \frac{R - \mu_R}{\sigma_R} = \frac{108 - 126.5}{15.23} = -1.215$

$-1.96 \quad \uparrow \quad 0 \qquad 1.96$
$\qquad -1.215$

5. continued

Do not reject the null hypothesis. There is not enough evidence to support the claim that there is a difference in scores.

7.

H_0: There is no difference in the stopping distances of the two types of automobiles. (claim)
H_1: There is a difference between the stopping distances of the two types of automobiles.

C. V. = ± 1.65

186	187	188	188	190	192	193	194	195	196
1	2	3.5	3.5	5	6	7	8	9	10
M	M	M	M	M	M	C	M	M	C

198	200	203	204	206	211	212	214	218	297
11	12	13	14	15	16	17	18	19	20
C	C	M	C	C	C	C	M	C	C

R = 69

$\mu_R = \frac{10(10 + 10 + 1)}{2} = 105$

$\sigma_R = \sqrt{\frac{10 \cdot 10(10 + 10 + 1)}{12}} = 13.23$

$z = \frac{69 - 105}{13.23} = -2.72$

$\uparrow \quad -1.96 \qquad 0 \qquad 1.96$
-2.72

Reject the null hypothesis. There is enough evidence to reject the claim that there is no difference in the stopping distances of the two types of automobiles.

9.

H_0: There is no difference in the number of hunting accidents in the two regions.
H_1: There is a difference in the number of hunting accidents in the two regions. (claim)
C. V. = ± 1.96

9. continued

2	3	5	5	6	7	8	8	9	10
1	2	3.5	3.5	5	6	7.5	7.5	9	10
E	E	E	E	E	E	E	W	W	W

11	11	11	13	13	14	15	17	17	21
12	12	12	14.5	14.5	16	17	18.5	18.5	20
W	W	E	E	W	E	W	W	W	W

R = 139

$$\mu_R = \frac{n_1(n_1 + n_2 + 1)}{2} = \frac{10(10 + 10 + 1)}{2} = 105$$

$$\sigma_R = \sqrt{\frac{n_1 \cdot n_2(n_1 + n_2 + 1)}{12}}$$

$$\sigma_R = \sqrt{\frac{10 \cdot 10(10 + 10 + 1)}{12}} = 13.23$$

$$Z = \frac{R - \mu_R}{\sigma_R} = \frac{139 - 105}{13.23} = 2.57$$

$-1.96 \qquad 0 \qquad 1.96 \qquad \uparrow$
2.57

Reject the null hypothesis. There is enough evidence to support the claim that there is a difference in the number of accidents in the two areas. The number of accidents may be related to the number of hunters in the two areas.

11.
H_0: There is no difference in time until pain is relieved.
H_1: There is a difference in time until pain is relieved. (claim)
C. V. = ± 1.96

7	9	9	10	10	11	11	12	12
1	2.5	2.5	4.5	4.5	6.5	6.5	8.5	8.5
N	N	N	N	N	A	N	A	N

13	14	14	15	15	16	17
10	11.5	11.5	13.5	13.5	15	16.5
N	A	N	A	A	N	A

17	18	20	20
16.5	18	19.5	19.5
A	A	A	A

R = 143.5

$$\mu_R = \frac{n_1(n_1 + n_2 + 1)}{2} = \frac{10(10 + 10 + 1)}{2}$$

11. continued

$$\mu_R = \frac{10(21)}{2} = 105$$

$$\sigma_R = \sqrt{\frac{n_1 \cdot n_2(n_1 + n_2 + 1)}{12}}$$

$$\sigma_R = \sqrt{\frac{10 \cdot 10(10 + 10 + 1)}{12}} = 13.23$$

$$Z = \frac{R - \mu_R}{\sigma_R} = \frac{143.5 - 105}{13.23} = 2.91$$

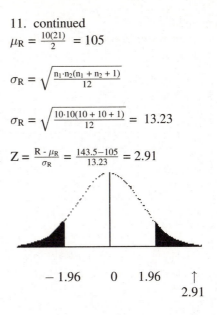

$-1.96 \qquad 0 \qquad 1.96 \qquad \uparrow$
2.91

Reject the null hypothesis. There is enough evidence to support the claim that there is a difference in time until pain is relieved.

EXERCISE SET 13-4

1.
The t-test for dependent samples

3.

B	A	B − A	|B − A|	Rank	Signed Rank
108	110	− 2	2	1	− 1
97	97	0			
115	103	12	12	4.5	4.5
162	168	− 6	6	2	− 2
156	143	13	13	6	6
105	112	− 7	7	3	− 3
153	141	12	12	4.5	4.5

Sum of the − ranks: $(-1) + (-2) + (-3) = -6$.
Sum of the + ranks: $4.5 + 6 + 4.5 = 15$
$w_s = 6$

5.
C. V. = 20 $\qquad$ $w_s = 18$
Since $18 \leq 20$, reject the null hypothesis.

7.
C. V. = 102 $\qquad$ $w_s = 102$
Since $102 = 102$, reject the null hypothesis.

9.
H_0: The human dose is less than or equal to the animal dose.

9. continued
H_1: The human dose is more than the animal dose. (claim)

H	A	H − A	|H − A|	Rank	Signed Rank
0.67	0.13	0.54	0.54	7	+ 7
0.64	0.18	0.46	0.46	6	+ 6
1.20	0.42	0.78	0.78	8	+ 8
0.51	0.25	0.26	0.26	4	+ 4
0.87	0.57	0.30	0.30	5	+ 5
0.74	0.57	0.17	0.17	3	+ 3
0.50	0.49	0.01	0.01	1	+ 1
1.22	1.28	− 0.06	0.06	2	− 2

Sum of the − ranks: − 2
Sum of the + ranks: + 34

$n = 8$ C. V. = 6
$w_s = |-2| = 2$
Since $2 < 6$, reject the null hypothesis. There is enough evidence to support the claim that the human dose costs more than the animal dose. One reason is that some people might not be inclined to pay a lot of money for their pets' medication.

11.
H_0: There is no difference in weights.
H_1: There is a difference in weights. (claim)

$n = 8$ $\alpha = 0.05$ C. V. = 4

B	A	B − A	|B − A|	Rank	Signed Rank
150	152	− 2	2	2.5	− 2.5
195	190	5	5	6.5	+ 6.5
188	185	3	3	4	+ 4
197	191	6	6	8	+ 8
204	200	4	4	5	+ 5
175	170	5	5	6.5	+ 6.5
160	162	− 2	2	2.5	− 2.5
180	179	1	1	1	+ 1

Sum of the − ranks: − 5
Sum of the + ranks: + 31
$w_s = 5$

Since $w_s >$ C. V., do not reject the null hypothesis. There is not enough evidence to conclude that there is a difference in weights.

13.
H_0: The prices of prescription drugs in the United States are greater than or equal to the prices in Canada.

13. continued
H_1: The drugs sold in Canada are cheaper. (claim)

$n = 10$ $\alpha = 0.05$ C. V. = 11

U. S.	C	US − C	|US − C|	Rank	Signed Rank
3.31	1.47	1.84	1.84	8	+ 8
2.27	1.07	1.20	1.20	4.5	+ 4.5
2.54	1.34	1.20	1.20	4.5	+ 4.5
3.13	1.34	1.79	1.79	7	+ 7
23.40	21.44	1.94	1.94	10	+ 10
3.16	1.47	1.69	1.69	6	+ 6
1.98	1.07	0.91	0.91	3	+ 3
5.27	3.39	1.88	1.88	9	+ 9
1.96	2.22	− 0.26	0.26	2	− 2
1.11	1.13	− 0.02	0.02	1	− 1

Sum of the + ranks:
$8 + 4.5 + 4.5 + 7 + 10 + 6 + 3 + 9 = 52$
Sum of the − ranks: $(-2) + (-1) = -3$

$w_s = |-3| = 3$

Since $3 < 11$, reject the null hypothesis. There is enough evidence to support the claim that the drugs are less expensive in Canada.

EXERCISE SET 13-5

1.
H_0: There is no difference in the number of calories each brand contains.
H_1: There is a difference in the number of calories each brand contains. (claim)
C. V. = 7.815 $\alpha = 0.05$ d. f. = 3

A	Rank	B	Rank	C	Rank	D	Rank
112	7	110	6	109	5	106	3
120	13	118	12	116	9.5	122	15
135	24	123	16	125	17.5	130	21.5
125	17.5	128	19.5	130	21.5	117	11
108	4	102	2	128	19.5	116	9.5
121	14	101	1	132	23	114	8
$R_1=$	79.5	$R_2=$	56.5	$R_3=$	96	$R_4=$	68

$$H = \frac{12}{N(N+1)} \left(\frac{R_1^2}{n_1} + \frac{R_2^2}{n_2} + \frac{R_3^2}{n_3} + \frac{R_4^2}{n_4} \right) - 3(N+1)$$

$$H = \frac{12}{12(24+1)} \left(\frac{79.5^2}{6} + \frac{56.5^2}{6} + \frac{96^2}{6} + \frac{68^2}{6} \right)$$

$$H = 2.842$$

1. continued

0   ↑   7.815
     2.842

Do not reject the null hypothesis. There is not enough evidence to support the claim that there is a difference in the calories.

3.
H_0: There is no difference in the prices of the three types of lawnmowers.
H_1: There is a difference in the prices of the three types of lawnmowers. (claim)
C. V. = 4.605   d. f. = 2 $\alpha = 0.10$

Self	Rank	Push	Rank	Electric	Rank
290	7	320	9	188	2
325	10	360	12	245	6
210	4	200	3	470	14
300	8	229	5	395	13
330	11	160	1		
R_1=	40	R_2=	30	R_3=	35

$$H = \frac{12}{N(N+1)} \left(\frac{R_1^2}{n_1} + \frac{R_2^2}{n_2} + \frac{R_3^2}{n_3} \right)$$

$$- 3(N+1)$$

$$H = \frac{12}{14(14+1)} \left(\frac{40^2}{5} + \frac{30^2}{5} + \frac{35^2}{4} \right) - 3(14+1)$$

$$H = \frac{12}{210} (320 + 180 + 306.25) - 3(15)$$

$$H = 1.07$$

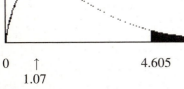

0   ↑   4.605
   1.07

Do not reject the null hypothesis. There is not enough evidence to support the claim that the prices are different. No, price is not a factor. Results are suspect since one sample is less than 5.

5.
H_0: There is no difference in unemployment benefits for the three states.
H_1: There is a difference in unemployment benefits for the three states. (claim)

C. V. = 5.991   d. f. = 2 $\alpha = 0.05$

FL	Rank	PA	Rank	ME	Rank
200	5	300	15	250	8
187	2	350	17	195	4
192	3	295	14	275	11
235	7	362	18	260	9.5
260	9.5	280	12	220	6
175	1	340	16	290	13
R_1=	27.5	R_2=	92	R_3=	51.5

$$H = \frac{12}{N(N+1)} \left(\frac{R_1^2}{n_1} + \frac{R_2^2}{n_2} + \frac{R_3^2}{n_3} + \frac{R_4^2}{n_4} \right) - 3(N+1)$$

$$= \frac{12}{18(18+1)} \left(\frac{27.5^2}{6} + \frac{92^2}{6} + \frac{51.5^2}{6} \right) - 3(18+1)$$

$$H = 12.43$$

0   5.991   ↑
                        12.43

Reject the null hypothesis. There is enough evidence to support the claim that there is a difference in unemployment benefits for the three states.

7.
H_0: There is no difference in spending between regions.
H_1: There is a difference in spending between regions.

C. V. = 5.991

E	Rank	M	Rank	W	Rank
6701	3	9854	15	7584	10
6708	4	8414	11	5474	1
9186	12	7279	7	6622	2
6786	5	7311	8	9673	14
9261	13	6947	6	7353	9
R_1=	37	R_2=	47	R_3=	36

$$H = \frac{12}{N(N+1)} \left(\frac{R_1^2}{n_1} + \frac{R_2^2}{n_2} + \frac{R_3^2}{n_3} + \frac{R_4^2}{n_4} \right) - 3(N+1)$$

7. continued

$$H = \frac{12}{15\,(15+1)}\left(\frac{37^2}{5} + \frac{47^2}{5} + \frac{36^2}{5}\right) - 3(15+1)$$

$$H = 0.74$$

$$\begin{array}{ccc} 0 & \uparrow & 5.991 \\ & 0.74 & \end{array}$$

Do not reject the null hypothesis. There is not enough evidence to conclude that there is a difference in spending.

9.

H_0: There is no difference in the number of crimes in the 5 precincts.

H_1: There is a difference in the number of crimes in the 5 precincts. (claim)

C. V. = 13.277   d. f. = 4 $\alpha = 0.01$

1	Rank	2	Rank	3	Rank
105	24	87	13	74	7.5
108	25	86	12	83	11
99	22	91	16	78	9
97	20	93	18	74	7.5
92	17	82	10	60	5
$R_1=$	108	$R_2=$	69	$R_3=$	40

4	Rank	5	Rank
56	3	103	23
43	1	98	21
52	2	94	19
58	4	89	15
62	6	88	14
$R_4=$	16	$R_5=$	92

$$H = \frac{12}{N(N+1)}\left(\frac{R_1^2}{n_1} + \frac{R_2^2}{n_2} + \frac{R_3^2}{n_3} + \frac{R_4^2}{n_4} + \frac{R_5^2}{n_5}\right) - 3(N+1)$$

$$H = \frac{12}{25(25+1)}\left(\frac{108^2}{5} + \frac{69^2}{5} + \frac{40^2}{5} + \frac{16^2}{5} + \frac{92^2}{5}\right) - 3(25+1)$$

$$H = 20.753$$

9. continued

$$\begin{array}{ccc} 0 & 13.277 & \uparrow \\ & & 20.753 \end{array}$$

Reject the null hypothesis. There is enough evidence to support the claim that there is a difference in the number of crimes for the precincts.

11.

H_0: There is no difference in speeds.

H_1: There is a difference in speeds.

d. f. = 2 $\alpha = 0.10$

C. V. = 5.991

Pred.	Rank	Deer	Rank	Dom.	Rank
70	15	50	12.5	47.5	11
50	12.5	35	5.5	39.35	7
43	10	32	4	35	5.5
42	9	30	2.5	30	2.5
40	8	61	14	11	1
$R_1=$	54.5	$R_2=$	38.5	$R_3=$	27

$$H = \frac{12}{N(N+1)}\left(\frac{R_1^2}{n_1} + \frac{R_2^2}{n_2} + \frac{R_3^2}{n_3}\right) - 3(N+1)$$

$$H = \frac{12}{12(15+1)}\left(\frac{54.5^2}{5} + \frac{38.5^2}{5} + \frac{27^2}{5}\right) - 3(15+1)$$

$$H = 3.815$$

Do not reject. There is not enough evidence to conclude there is a difference in speeds.

EXERCISE SET 13-6

1.
0.716

3.
0.648

5.
H_0: $\rho = 0$
H_1: $\rho \neq 0$
C. V. = ± 0.786 n = 7 $\alpha = 0.05$

5. continued

Grade 4	R_1	Grade 8	R_2	$R_1 - R_2$	d^2
89	6.5	81	7	-0.5	0.25
84	4	75	4	0	0
80	2.5	66	2	0.5	0.25
89	6.5	76	5	1.5	2.25
88	5	80	6	-1	1
77	1	59	1	0	0
80	2.5	74	3	-0.5	0.25
				$\sum d^2 =$	4

$$r_s = 1 - \frac{6 \cdot \sum d^2}{n(n^2-1)} = 1 - \frac{6 \cdot 4}{7(7^2-1)}$$

$$r_s = 0.929$$

Reject the null hypothesis. There is a significant relationship between the mathematics achievement scores for the two grades.

7.

H_0: $\rho = 0$

H_1: $\rho \neq 0$

C. V. $= \pm 0.700$ $n = 9$ $\alpha = 0.05$

Releases	R_1	Receipts	R_2	$R_1 - R_2$	d^2
361	9	3844	9	0	0
270	7	1967	8	-1	1
306	8	1371	7	1	1
22	5	1064	6	-1	1
35	6	667	5	1	1
10	2	241	4	-2	4
8	1	188	3	-2	4
12	3	154	2	1	1
21	4	125	1	3	9
				$\sum d^2 =$	22

$$r_s = 1 - \frac{6 \cdot \sum d^2}{n(n^2-1)}$$

$$r_s = 1 - \frac{6 \cdot 22}{9(9^2-1)} = 0.817$$

Reject. There is a significant relationship between the number of new releases and gross receipts.

9.

H_0: $\rho = 0$

H_1: $\rho \neq 0$

C. V. $= \pm 0.738$

9. continued

Calories	R_1	Cholesterol	R_2	$R_1 - R_2$	d^2
580	7.5	205	3	4.5	20.25
580	7.5	225	6	1.5	2.25
270	1	285	8	-7	49
470	6	270	7	-1	1
420	4	185	1.5	2.5	6.25
415	3	215	4	-1	1
330	2	185	1.5	0.5	0.25
430	5	220	5	0	0
				$\sum d^2 =$	80

$$r_s = 1 - \frac{6 \cdot 80}{8(8^2-1)} = 1 - \frac{480}{504} = 0.048$$

Do not reject the null hypothesis. There is not enought evidence to say that a significant relationship exists between the rankings.

11.

H_0: $\rho = 0$

H_1: $\rho \neq 0$

C. V. $= \pm 0.700$

Agency	R_1	Station	R_2	$R_1 - R_2$	d^2
5.12	1.5	2.09	3	-1.5	2.25
5.27	5	1.96	2	3.0	9.00
5.29	6	2.29	9	-3.0	9.00
5.18	4	1.94	1	3.0	9.00
5.59	9	2.20	7.5	1.5	2.25
5.30	7	2.20	7.5	-0.5	0.25
5.83	10	2.40	10	0	0
5.46	8	2.12	5	3.0	9.00
5.12	1.5	2.15	6	-4.5	20.25
5.15	3	2.11	4	-1.0	1.00
				$\sum d^2 =$	62.00

$$r_s = 1 - \frac{6\sum d^2}{n(n^2-1)} = 1 - \frac{6 \cdot 62}{10(10^2-1)} = 0.624$$

Do not reject the null hypothesis. There is no significant linear relationship between gasoline prices at a car rental agency and prices at a gas station. One would wonder how the car rental agencies determine their prices.

13.

H_0: $\rho = 0$

H_1: $\rho \neq 0$

C. V. $= \pm 0.900$

13. continued

Students	R_1	Cost	R_2	$R_1 - R_2$	d^2
10	3	7200	2	1	1
6	1	9393	5	-4	16
17	5	7385	3	2	4
8	2	4500	1	1	1
11	4	8203	4	0	0
				$\sum d^2 =$	22

$$r_s = 1 - \frac{6\sum d^2}{n(n^2-1)} = 1 - \frac{6 \cdot 22}{5(5^2-1)}$$

$$r_s = -0.10$$

Do not reject the null hypothesis. There is no significant relationship exists between the number of cyber school students and the cost per pupil. In this case, the cost per pupil is different in each district.

15.
H_0: The occurrance of cavities is random.
H_1: The null hypothesis is not true.

The median of the data set is two. Using A = above and B = below, the runs (going across) are shown:

B AA B AAA B A BB AAAA B A B A B A B A B AAA B A BB

There are 21 runs. The expected number of runs is between 10 and 22. Therefore, the null hypothesis should not be rejected. The number of cavities occurs at random.

17.
H_0: The purchases of soft drinks occur at random.
H_1: The null hypothesis is not true.

PP CCC P C PP C PPP C P C P CCCC P CC PPPP C

There are 16 runs and this value is between 9 and 22. Hence, do not reject the null hypothesis. The purchases occur at random.

19.
H_0: The seating occurs at random.
H_1: The null hypothesis is not true.

BB GG BB G BBBBB G BB GG BBBB GGGG B G BBB GG

There are 14 runs and since this is between 10 and 23, the null hypothesis is not rejected. The seating occurs at random.

21.
H_0: The number of absences of employees occurs at random.
H_1: The null hypothesis is not true.

The median of the data is 12. Using A = above and B = below, the runs are shown as follows:

A B AAAAAAA BBBBBBBBB AAAAAA BBBB

There are six runs. The expected number of runs is between 9 and 21, hence the null hypothesis is rejected since six is not between 9 and 21. The number of absences do not occur at random.

23.
Answers will vary.

25.
$r = \frac{\pm z}{\sqrt{n-1}} = \frac{\pm 2.58}{\sqrt{30-1}} = \pm 0.479$

27.
$r = \frac{\pm z}{\sqrt{n-1}} = \frac{\pm 1.65}{\sqrt{60-1}} = \pm 0.215$

REVIEW EXERCISES - CHAPTER 13

1.
H_0: median $= 36$
H_1: median $\neq 36$

C. V. $= \pm 1.96$ $\alpha = 0.05$

$$Z = \frac{(13 + 0.5) - \left(\frac{30}{2}\right)}{\frac{\sqrt{30}}{2}} = -0.548$$

-1.96 ↑ 0   1.96
-0.548

Do not reject. There is not enough evidence to conclude that the median differs from 36.

3.
H_0: The special diet has no effect on weight.
H_1: The diet increases weight.   (claim)

$R^+ = 35, R^- = 1$
C. V. $= 0$

3. continued
Test value = 1

Do not reject the null hypothesis. There is not enough evidence to conclude a difference in prices. Comments: Examine what affects the result of this test.

5.
H_0: There is no difference in the amount of hours worked.
H_1: There is a difference in the amount of hours worked.

C. V. = ± 1.645 at $\alpha = 0.10$

12	15	17	18	18	19	19	20	20
1	2	3	4.5	4.5	6.5	6.5	8.5	8.5
A	A	A	A	L	A	L	A	L

21	22	22	24	24	25	25	26
10	11.5	11.5	13.5	13.5	15.5	15.5	17
L	A	L	A	L	L	A	L

28	30	31	35
18	19	20	21
L	A	L	L

R = 85

$$\mu_R = \frac{n_1(n_1 + n_2 + 1)}{2} = \frac{10(10 + 11 + 1)}{2} = 110$$

$$\sigma_R = \sqrt{\frac{n_1 n_2 (n_1 + n_2 + 1)}{12}}$$

$$\sigma_R = \sqrt{\frac{10 \cdot 11(10 + 11 + 1)}{12}} = 14.201$$

$$Z = \frac{R - \mu_R}{\sigma_R} = \frac{85 - 110}{14.201} = -1.76$$

$\uparrow -1.65 \quad 0 \quad 1.65$
-1.76

Reject the null hypothesis. There is enough evidence to conclude a difference in the hours worked.

At $\alpha = 0.05$, C. V. = ± 1.96. The decision would be do not reject the null hypothesis.

7.
H_0: There is no difference in the amount spent.
H_1: There is a difference in the amount spent.

B	A	B − A	|B − A|	Rank	Signed Rank
7	6	1	1	1	1
5.5	10	− 4.5	4.5	7	− 7
4.5	7	− 2.5	2.5	6	− 6
10	12	− 2	2	4	− 4
6.75	8.5	− 1.75	1.75	2	− 2
5	7	− 2	2	4	− 4
6	8	− 2	2	4	− 4

Sum of the + ranks: 1
Sum of the − ranks: − 27
$w_s = 1$
C. V. = 2 $\quad \alpha = 0.05 \quad n = 7$

Reject the null hypothesis. There is enough evidence to conclude that there is a difference in amount spent.

9.
H_0: There is no difference in beach temperatures.
H_1: There is a difference in beach temperatures.

C. V. = 7.815

S. Pac.	Rank	W. Gulf	Rank	E. Gulf	Rank
67	4	86	21	87	25
68	5	86	21	87	25
66	3	84	13	86	21
69	6	85	16.5	86	21
63	2	79	8	85	16.5
62	1	85	16.5	84	13
				85	16.5
$R_1=$	21	$R_2=$	96	$R_3=$	138

S. Atl.	Rank
76	7
81	10
82	11
84	13
80	9
86	21
87	25
$R_4=$	96

$$H = \frac{12}{N(N+1)} \left(\frac{R_1^2}{n_1} + \frac{R_2^2}{n_2} + \frac{R_3^2}{n_3} \right) - 3(N+1)$$

$$H = \frac{12}{26(26 + 1)} \left(\frac{21^2}{6} + \frac{96^2}{6} + \frac{138^2}{7} + \frac{96^2}{7} \right)$$

$$- 3(26 + 1)$$

9. continued
H = 15.524

0   7.815   ↑
                               15.524

Reject the null hypothesis. There is enough evidence to conclude a difference in beach temperatures.

Without the Southern Pacific:

C. V. = 5.991 at $\alpha = 0.05$

$$H = \frac{12}{20(20+1)}\left(\frac{60^2}{6} + \frac{96^2}{7} + \frac{54^2}{7}\right)$$

$$- 3(20+1)$$

H = 3.661

0   ↑   5.991
          3.661

Do not reject the null hypothesis. There is not enough evidence to conclude a difference in the temperatures. The conclusion is not the same without the Southern Pacific temperatures.

11.

Last Year	R_1	This Year	R_2	$R_1 - R_2$	d^2
28.9	10	26.6	10	0	0
26.4	9	20.5	9	0	0
20.8	6	20.2	8	− 2	4
25.0	8	19.1	7	1	1
21.0	7	18.9	6	1	1
19.2	5	17.8	5	0	0
13.7	1	16.8	4	− 3	9
18.8	4	16.7	3	1	1
16.8	3	16.0	2	1	1
15.3	2	15.8	1	1	1
				$\sum d^2 =$	18

$$r_s = 1 - \frac{6\sum d^2}{n(n^2-1)} = 1 - \frac{6(18)}{10(99)} = 0.891$$

11. continued
H_0: $\rho = 0$
H_1: $\rho \neq 0$

C. V. = ± 0.648

Reject the null hypothesis. There is a significant relationship in the average number of people who are watching the television shows for both years.

13.
H_0: The grades of students who finish the exam occur at random.
H_1: The null hypothesis is not true.

The median grade is 73. Using A = above and B = below, the runs are:

AAAA B AAAA BBBB AAAAA BB A BBBBBBB

Since there are eight runs and this does not fall between 9 and 21, the null hypothesis is rejected. The grades do not occur at random.

CHAPTER 13 QUIZ

1. False
2. False, they are less sensitive.
3. True
4. True
5. a
6. c
7. d
8. b
9. Non-parametric
10. Nominal, ordinal
11. Sign
12. Sensitive

13. H_0: Median = \$177,500
H_1: Median $\neq$ \$177,500 (claim)
There are three $-$ signs. Do not reject since 3 is greater than the critical value of 2. There is not enough evidence to support the claim that the median is not \$177,500.

14. H_0: Median = 1200 (claim)
H_1: Median $\neq$ 1200
There are ten $-$ signs. Do not reject since 10 is greater than the critical value 6. There is not enough evidence to reject the claim that the median is 1,200.

15. H_0: There will be no change in the weight of the turkeys after the special diet.
H_1: The turkeys will weigh more after the special diet. (claim)
There is one + sign. Reject the null hypothesis. There is enough evidence to support the claim that the turkeys gained weight on the special diet.

16. H_0: There is not difference in the amounts of money received by the teams.
H_1: There is a difference in the amounts of money received by the teams. (claim)
C. V. = ± 1.96
z = $- 0.79$
Do not reject the null hypothesis. There is not enough evidence to support the claim that the distributions are different.

17. H_0: The distributions are the same.
H_1: The distributions are different. (claim)
C. V. = ± 1.65
z = $- 0.144$
Do not reject the null hypothesis. There is not enough evidence to support the claim that the distributions are different.

18. H_0: There is no difference in the GPAs before and after the workshop.
H_1: There is a difference in GPAs before and after the workshop. (claim)
C. V. = 2   Test statistic = 0
Reject the null hypothesis. There is enough evidence to support the claim that there is a difference in the GPAs of the students.

19. H_0: There is no difference in the breaking strengths of the tapes.
H_1: There is a difference in breaking strengths. (claim)
H = 11.795
C. V. = 5.991
Reject the null hypothesis. There is enough evidence to support the claim that there is a difference in the breaking strengths of the tapes.

20. H_0: There is no difference in reaction times.
H_1: There is a difference in reaction times. (claim)
H = 6.9
$0.025 <$ P-value < 0.05 (0.032)
Reject the null hypothesis. There is enough evidence to support the claim that there is a difference in the reaction times of the monkeys.

21. H_0: $\rho = 0$
H_1: $\rho \neq 0$

21. continued
C. V. = ± 0.600
$r_s = 0.683$
Reject the null hypothesis. There is a significant relationship between the drug prices.

22. H_0: $\rho = 0$
H_1: $\rho \neq 0$
$r_s = 0.943$
C. V. = ± 0.829
Reject the null hypothesis. There is a significant relationship between the amount of money spent on Head Start and the number of students enrolled in the program.

23. H_0: The gender of babies occurs at random.
H_1: The null hypothesis is false.
$\alpha = 0.05$ C. V. = 8, 19
There are 10 runs, which is between 8 and 19. Do not reject the null hypothesis. There is not enough evidence to reject the null hypothesis that the gender occurs at random.

24. H_0: There is no difference in output ratings.
H_1: There is a difference in output ratings after reconditioning. (claim)
$\alpha = 0.05$ n = 9   C. V. = 6
Test statistic = 0
Do not reject the null hypothesis. There is not enough evidence to support the claim that there is a difference in output ratings before and after reconditioning.

25. H_0: The numbers occur at random
H_1: The null hypothesis is false.
$\alpha = 0.05$ C. V. = 9, 21
The median number is 538.
There are 16 runs, reading from left to right, and since this is between 9 and 21, the null hypothesis is not rejected. There is not enough evidence to reject the null hypothesis that the numbers occur at random.

EXERCISE SET 14-1

1.
Random, systematic, stratified, cluster.

3.
A sample must be randomly selected.

5.
Talking to people on the street, calling people on the phone, and asking one's friends are three incorrect ways of obtaining a sample.

7.
Random sampling has the advantage that each unit of the population has an equal chance of being selected. One disadvantage is that the units of the population must be numbered, and if the population is large this could be somewhat time consuming.

9.
An advantage of stratified sampling is that it ensures representation for the groups used in stratification; however, it is virtually impossible to stratify the population so that all groups could be represented.

11 through 15.
Answers will vary.

EXERCISE SET 14-2

1.
Flaw - biased; it's confusing.

3. Flaw - the question is too broad.

5.
Flaw - confusing wording. The question could be worded: "How many hours did you study for this exam?"

7.
This question has confusing wording. Change the question to read: "If a plane were to crash on the border of New York and New Jersey, where should the victims be buried?"

9.
Answers will vary.

EXERCISE SET 14-3

1.
Simulation involves setting up probability experiments that mimic the behavior of real life events.

3.
John Van Neumann and Stanislaw Ulam.

5.
The steps are:
1. List all possible outcomes.
2. Determine the probability of each outcome.
3. Set up a correspondence between the outcomes and the random numbers.
4. Conduct the experiment using random numbers.
5. Repeat the experiment and tally the outcomes.
6. Compute any statistics and state the conclusions.

7.
When the repetitions increase there is a higher probability that the simulation will yield more precise answers.

9.
Use three-digit random numbers; numbers 001 through 681 mean that the mother is in the labor force.

11.
Select 100 two-digit random numbers. Numbers 01 through 35 mean the household has at least one set with permium cable service. Numbers 36 to 100 mean the household does not have the service.

13.
Let an odd number represent heads and an even number represent tails. Then each person selects a digit at random.

15 through 23.
Answers will vary.

REVIEW EXERCISES - CHAPTER 14

1 - 3.
Answers will vary.

5 - 7.

Answers will vary. The range of the data is fairly narrow, and the data are not ranked in any way. Regardless of the method of sampling used, the sample means should be about the same.

9.

Use one-digit random numbers 1 through 4 to represent a strikeout and 5 through 9 and 0 to represent anything other than a strikeout.

11.

In this case, a one-digit random number is selected. Numbers 1 through 6 represent the numbers on the face. Ignore 7, 8, 9, and 0 and select another number.

13.

Let the digits 1 – 3 represent "rock"
Let the digits 4 – 6 represent "paper"
Let the digits 7 – 9 represent "scissors"
Omit 0.

15 - 17.

Answers will vary.

19.

This is a biased question. Change the question to read: "Have you ever driven through a red light?"

21.

This is a double-barreled question. Change the question to read: "Do you think all automobiles should have heavy-duty bumpers?"

CHAPTER 14 QUIZ

1. True
2. True
3. False, only random numbers generated by a random number table are random.
4. True
5. a.
6. c.
7. c.
8. larger
9. biased
10. cluster
11 through 14. Answers will vary.

15. Use two-digit random numbers: 01 through 45 constitute a win. Any other two-digit number means the player loses.

16. Use two-digit random numbers: 01 - 05 constitute a cancellation. Any other two-digit random number means the person shows up.

17. Use two-digit random numbers 01 through 13 to represent the 13 cards in hearts. The random numbers 14 through 26 represent the 13 cards in diamonds. The random numbers 27 through 39 represent the 13 spades, and 40 through 52 represent the 13 clubs. Any number over 52 is ignored.

18. Use two-digit random numbers to represent the spots on the face of the dice. Ignore any two-digit random numbers with 7, 8, 9, or 0. For cards, use two-digit random numbers between 01 and 13.

19. Use two-digit random numbers. The first digit represents the first player, and the second digit represents the second player. If both numbers are odd or even, player 1 wins. If a digit is odd and the other digit is even, player 2 wins.

20 through 24. Answers will vary.

A-1

A-1. $9! = 9 \cdot 8 \cdot 7 \cdot 6 \cdot 5 \cdot 4 \cdot 3 \cdot 2 \cdot 1 = 362,880$

A-2. $7! = 7 \cdot 6 \cdot 5 \cdot 4 \cdot 3 \cdot 2 \cdot 1 = 5040$

A-3. $5! = 5 \cdot 4 \cdot 3 \cdot 2 \cdot 1 = 120$

A-4. $0! = 1$

A-5. $1! = 1$

A-6. $3! = 3 \cdot 2 \cdot 1 = 6$

A-7. $\frac{12!}{9!} = \frac{12 \cdot 11 \cdot 10 \cdot 9!}{9!} = 1320$

A-8. $\frac{10!}{2!} = \frac{10 \cdot 9 \cdot 8 \cdot 7 \cdot 6 \cdot 5 \cdot 4 \cdot 3 \cdot 2!}{2!}$

$= 1,814,400$

A-9. $\frac{5!}{3!} = \frac{5 \cdot 4 \cdot 3!}{3!} = 20$

A-10. $\frac{11!}{7!} = \frac{11 \cdot 10 \cdot 9 \cdot 8 \cdot 7!}{7!} = 7920$

A-11. $\frac{9!}{(4!)(5!)} = \frac{9 \cdot 8 \cdot 7 \cdot 6 \cdot 5!}{4 \cdot 3 \cdot 2 \cdot 1 \cdot 5!} = 126$

A-12. $\frac{10!}{(7!)(3!)} = \frac{10 \cdot 9 \cdot 8 \cdot 7!}{3 \cdot 2 \cdot 1 \cdot 7!} = 120$

A-13. $\frac{8!}{4!4!} = \frac{8 \cdot 7 \cdot 6 \cdot 5 \cdot 4!}{4 \cdot 3 \cdot 2 \cdot 1 \cdot 4!} = 70$

A-14. $\frac{15!}{12!3!} = \frac{15 \cdot 14 \cdot 13 \cdot 12!}{3 \cdot 2 \cdot 1 \cdot 12!} = 455$

A-15. $\frac{10!}{(10!)(0!)} = \frac{10!}{10! \cdot 1} = 1$

A-16. $\frac{5!}{3!2!1!} = \frac{5 \cdot 4 \cdot 3!}{3! \cdot 2 \cdot 1 \cdot 1} = 10$

A-17. $\frac{8!}{3!3!2!} = \frac{8 \cdot 7 \cdot 6 \cdot 5 \cdot 4 \cdot 3!}{3! \cdot 3 \cdot 2 \cdot 1 \cdot 2 \cdot 1} = 560$

A-18. $\frac{11!}{7!2!2!} = \frac{11 \cdot 10 \cdot 9 \cdot 8 \cdot 7!}{7! \cdot 2 \cdot 1 \cdot 2 \cdot 1} = 1980$

A-19. $\frac{10!}{3!2!5!} = \frac{10 \cdot 9 \cdot 8 \cdot 7 \cdot 6 \cdot 5!}{3 \cdot 2 \cdot 1 \cdot 2 \cdot 1 \cdot 5!} = 2520$

A-20. $\frac{6!}{2!2!2!} = \frac{6 \cdot 5 \cdot 4 \cdot 3 \cdot 2!}{2 \cdot 1 \cdot 2 \cdot 1 \cdot 2!} = 90$

A-2

A-21.

X	X^2	$X - \overline{X}$	$(X - \overline{X})^2$
9	81	-3.1	9.61
17	289	4.9	24.01
32	1024	19.9	396.01
16	256	3.9	15.21
8	64	-4.1	16.81
2	4	-10.1	102.01
9	81	-3.1	9.61
7	49	-5.1	26.01
3	9	-9.1	82.81
$\underline{18}$	$\underline{324}$	$\underline{5.9}$	$\underline{34.81}$
121	2181		716.9

$\sum X = 121 \quad \overline{X} = \frac{121}{10} = 12.1 \quad \sum X^2 = 2181$

$(\sum X)^2 = 121^2 = 14641 \quad \sum (X-\overline{X})^2 = 716.9$

A-22.

X	X^2	$X - \overline{X}$	$(X - \overline{X})^2$
4	16	-3	9
12	144	5	25
9	81	2	4
13	169	6	36
0	0	-7	49
6	36	-1	1
2	4	-5	25
$\underline{10}$	$\underline{100}$	$\underline{3}$	$\underline{9}$
56	550		158

$\sum X = 56 \quad \overline{X} = \frac{56}{8} = 7 \quad \sum X^2 = 550$

$(\sum X)^2 = 56^2 = 3136 \quad \sum (X-\overline{X})^2 = 158$

A-23.

X	X^2	$X - \overline{X}$	$(X - \overline{X})^2$
5	25	-1.4	1.96
12	144	5.6	31.36
8	64	1.6	2.56
3	9	-3.4	11.56
$\underline{4}$	$\underline{16}$	-2.4	$\underline{5.76}$
32	258		53.20

$\sum X = 32 \quad \overline{X} = \frac{32}{5} = 6.4 \quad \sum X^2 = 258$

$(\sum X)^2 = 32^2 = 1024 \quad \sum (X-\overline{X})^2 = 53.2$

A-24.

X	X^2	$X - \overline{X}$	$(X - \overline{X})^2$
6	36	-12.75	163.5625
2	4	-16.75	280.5625
18	324	-0.75	0.5625
30	900	11.25	126.5625
31	961	12.25	150.0625
42	1764	23.25	540.5625
16	256	-2.75	7.5625
5	25	-13.75	189.0625
150	4270		1457.5000

$\sum X = 150$ $\overline{X} = \frac{150}{8} = 18.75$ $\sum X^2 = 4270$

$(\sum X)^2 = 150^2 = 22500$ $\sum (X-\overline{X})^2 = 1457.5$

A-25.

X	X^2	$X - \overline{X}$	$(X - \overline{X})^2$
80	6400	14.4	207.36
76	5776	10.4	108.16
42	1764	-23.6	556.96
53	2809	-12.6	158.76
77	5929	11.4	129.96
328	22678		1161.20

$\sum X = 328$ $\overline{X} = \frac{328}{5} = 65.6$ $\sum X^2 = 22678$

$(\sum X)^2 = 328^2 = 107584$ $\sum (X-\overline{X})^2 = 1161.2$

A-26.

X	X^2	$X - \overline{X}$	$(X - \overline{X})^2$
123	15129	-15.17	230.1289
132	17424	-6.17	38.0689
216	46656	77.83	6057.5089
98	9604	-40.17	1613.6289
146	21316	7.83	61.3089
114	12996	-24.17	584.1889
829	123125		8584.8334

$\sum X = 829$ $\overline{X} = \frac{829}{6} = 138.17$

$\sum X^2 = 123125$ $(\sum X)^2 = 829^2 = 687241$

$\sum (X-\overline{X})^2 = 8584.8334$

A-27.

X	X^2	$X - \overline{X}$	$(X - \overline{X})^2$
53	2809	-16.3	265.69
72	5184	2.7	7.29
81	6561	11.7	136.89
42	1764	-27.3	745.29
63	3969	-6.3	39.69
71	5041	1.7	2.89
73	5329	3.7	13.69
85	7225	15.7	246.49
98	9604	28.7	823.69
55	3025	-14.3	204.49
693	50511		2486.10

$\sum X = 693$ $\overline{X} = \frac{693}{10} = 69.3$ $\sum X^2 = 50511$

$(\sum X)^2 = 693^2 = 480249$ $\sum (X-\overline{X})^2 = 2486.1$

A-28.

X	X^2	$X - \overline{X}$	$(X - \overline{X})^2$
43	1849	-38.8	1505.44
32	1024	-49.8	2480.04
116	13456	34.2	1169.64
98	9604	16.2	262.44
120	14400	38.2	1459.24
409	40333		6876.80

$\sum X = 409$ $\overline{X} = \frac{409}{5} = 81.8$ $\sum X^2 = 40333$

$(\sum X)^2 = 409^2 = 167281$ $\sum (X-\overline{X})^2 = 6876.8$

A-29.

X	X^2	$X - \overline{X}$	$(X - \overline{X})^2$
12	144	-41	1681
52	2704	-1	1
36	1296	-17	289
81	6561	28	784
63	3969	10	100
74	5476	21	441
318	20150		3296

$\sum X = 318$ $\overline{X} = \frac{318}{6} = 53$ $\sum X^2 = 20150$

$(\sum X)^2 = 318^2 = 101124$ $\sum (X-\overline{X})^2 = 3296$

A-30.

X	X^2	$X - \overline{X}$	$(X - \overline{X})^2$
-9	81	-5.67	32.1489
-12	144	-8.67	75.1689
18	324	21.33	454.9689
0	0	3.33	11.0889
-2	4	1.33	1.7689
$=15$	225	-11.67	136.1889
-20	778		711.3334

$\overline{X} = \frac{-20}{6} = -3.33 \quad (\sum X)^2 = -20^2 = 400$

A-3

A-31.

A-32.

A-33.

A-34.

A-35.

A-36.

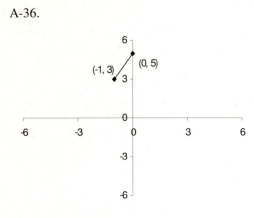

Two points are: (0, 5) and (-1, 3).

A-37.

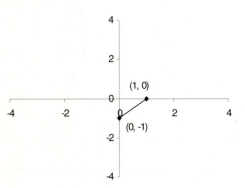

Two points are: (1, 0) and (0, -1).

A-38.

Two points are: (1, 7) and (0, 3).

A-39

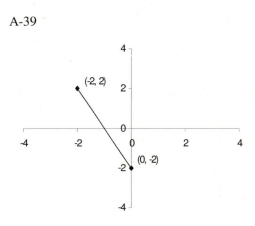

Two points are: (-2, 2) and (0, -2).

A-40.

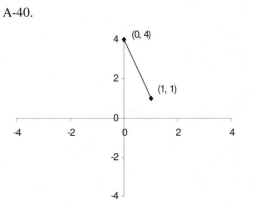

Two points are: (1, 1) and (0, 4)

Notes

Notes

Notes

Notes

Notes